SOLID EDGE EXERCISES

200 PRACTICE DRAWINGS

SACHIDANAND JHA

Dear Reader,

Thank you for choosing **SIEMENS SOLID EDGE EXERCISES** book. This book is part of a family of premium-quality CADIN360 books, all of which are written by Outstanding author who combine practical experience with a gift for teaching.

CADIN360 was founded in 2016. More than 3 years later, we're still committed to producing consistently exceptional books. With each of our titles, we're working hard to set a new standard for the industry. From the paper we print on, to the authors we work with, our goal is to bring you the best books available.

I hope you see all that reflected in these pages. I'd be very interested to hear your comments and get your feedback on how we're doing. Feel free to let me know what you think about this or any other CADIN360 book by sending me an email at contactus@cadin360.com.

If you think you've found a technical error in this book, please visit
https://cadin360.com/contact-us/.
Customer feedback is critical to our efforts at CADIN360.

Best regards,

Sachidanand Jha
Founder & CEO, CADIN360

SIEMENS SOLID EDGE EXERCISES

Published by
CADIN360
cadin360.com

Limit of Liability/Disclaimer of Warranty:

Examination Copies

Electronic Files

Disclaimer:

Preface

SIEMENS SOLID EDGE EXERCISES

- ❖ This book contain 200 CAD practice exercises and drawings.

- ❖ This book does not provide step by step tutorial to design 3D models.

- ❖ S.I Unit is used.

- ❖ Predominantly used Third Angle Projection.

- ❖ This book is for **SIEMENS SOLID EDGE** and Other Feature-Based Modeling Software such as Inventor, SolidWorks, NX, AutoCAD, PTC Creo etc.

- ❖ It is intended to provide Drafters, Designers and Engineers with enough 3D CAD exercises for practice on **SIEMENS SOLID EDGE**.

- ❖ It includes almost all types of exercises that are necessary to provide, clear, concise and systematic information required on industrial machine part drawings.

- ❖ Third Angle Projection is intentionally used to familiarize Drafters, Designers and Engineers in Third Angle Projection to meet the expectation of world wide Engineering drawing print.

- ❖ Clear and well drafted drawing help easy understanding of the design.

- ❖ This book is for Beginner, Intermediate and Advance CAD users.

- ❖ These exercises are from Basics to Advance level.

- ❖ Each exercises can be assigned and designed separately.

- ❖ No Exercise is a prerequisite for another. All dimensions are in mm.

- ❖ Note: Assume any missing dimensions.

EX-01
⌀80
3 HOLES ⌀10
DRILLED THROUGH
28
28
A
28 A
28
10
5
SECTION A-A

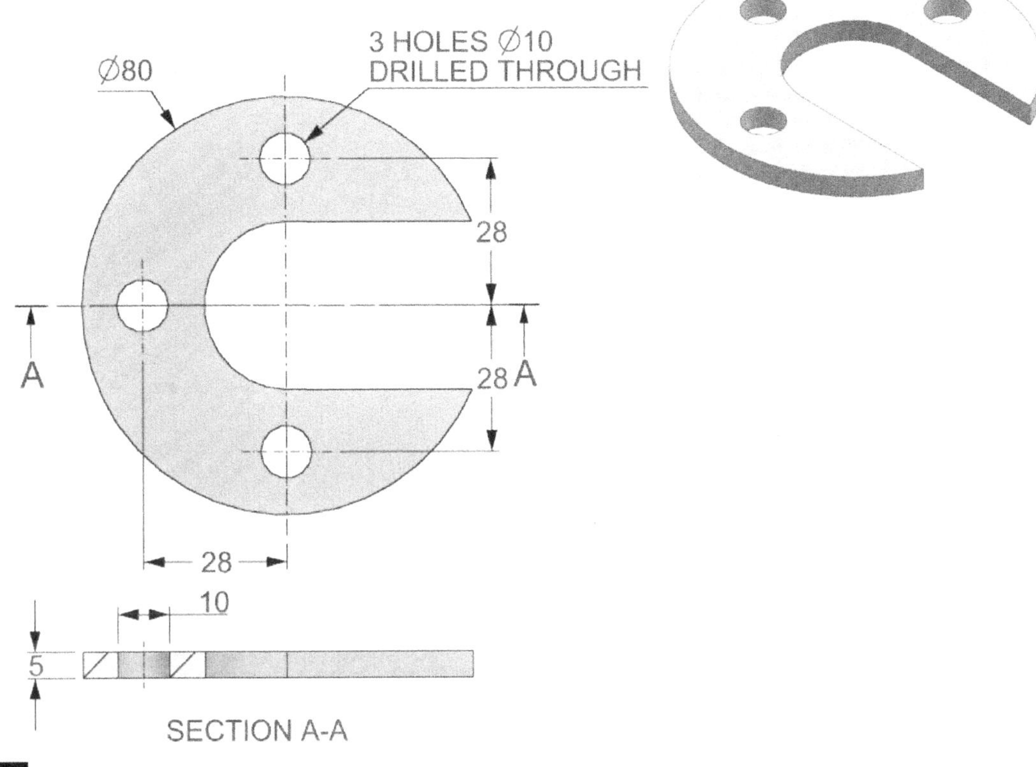

EX-02
100
20
2X R30
2X R15
120
60
30
70
20
50
20
100
20
80
2X R20
2X R6
30
20
20
30
120
P-01

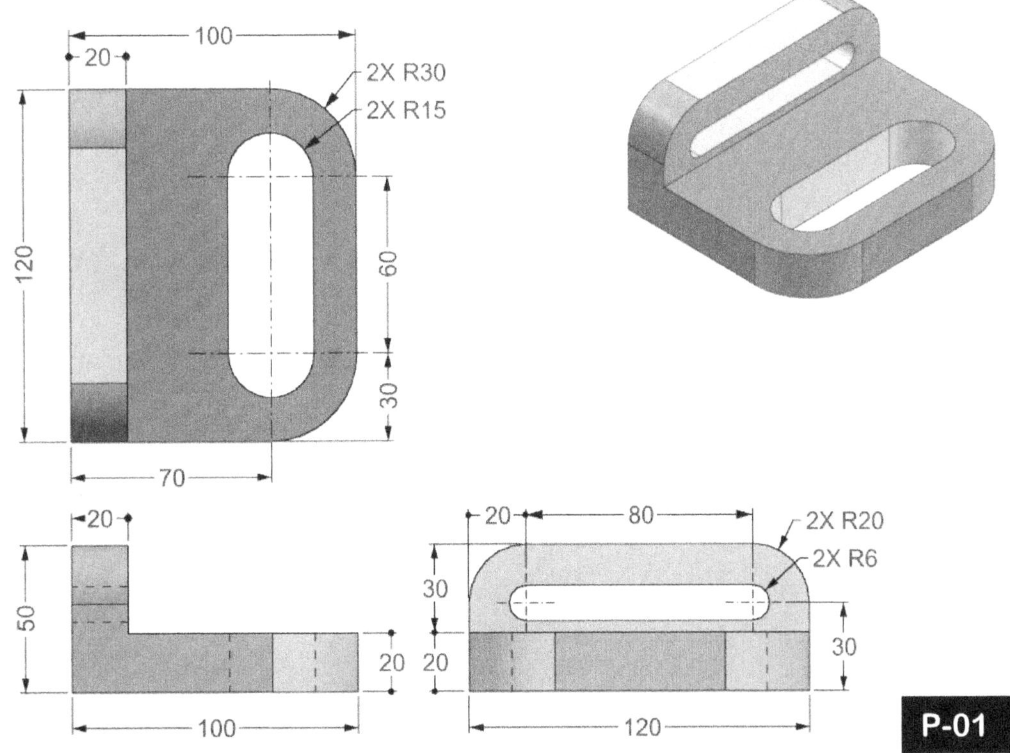

EX-03
25
Ø50
20
60
100
50
R25
Ø40
45
20
20
100
20
70
45
20
60

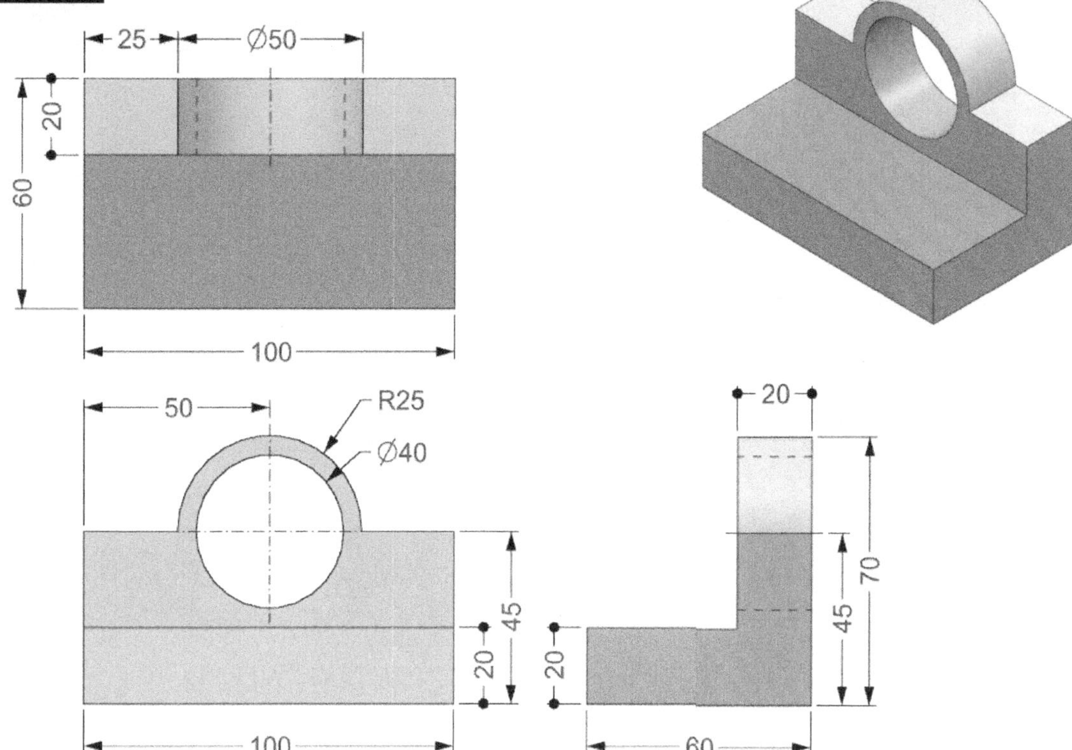

EX-04
Ø40
30
30
Ø20

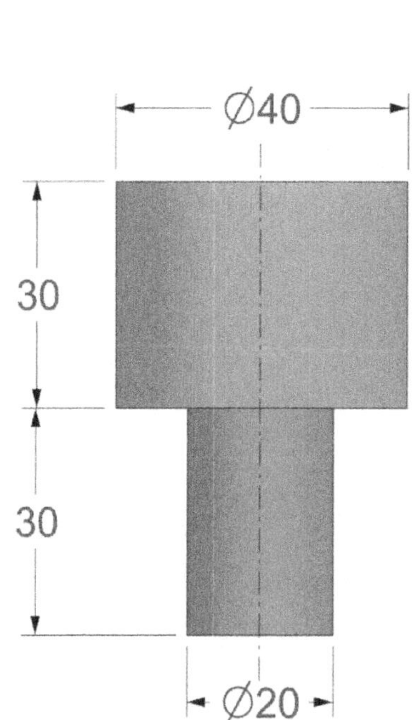

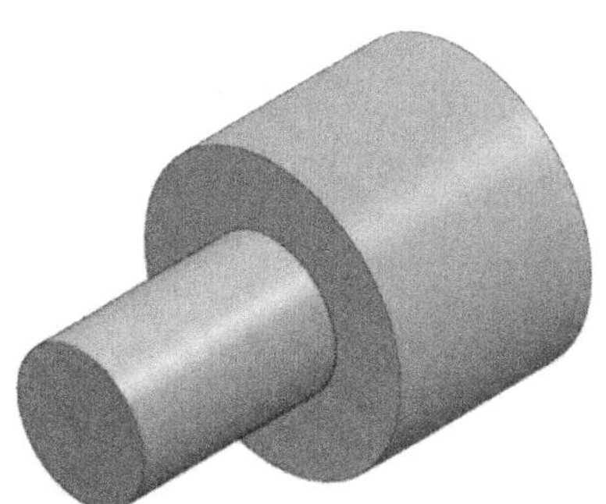

P-02

EX-05

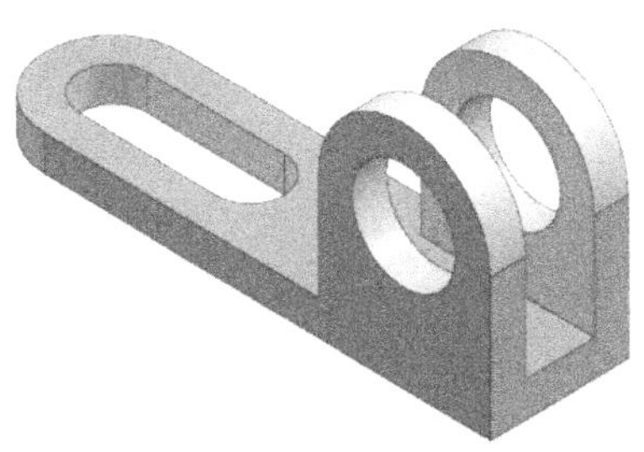

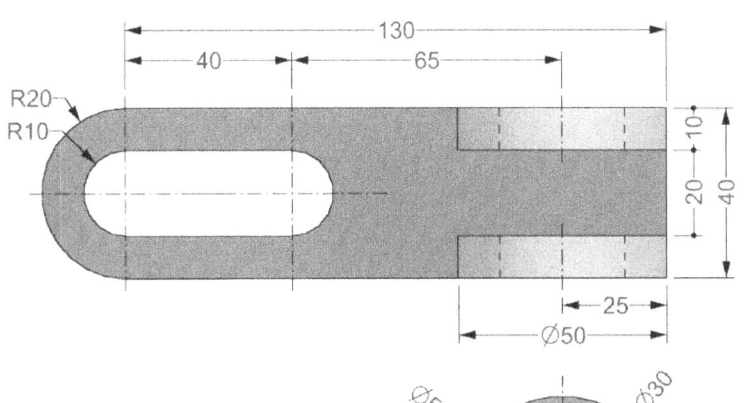

130
40
65
R20
R10
10
20
40
25
Ø50

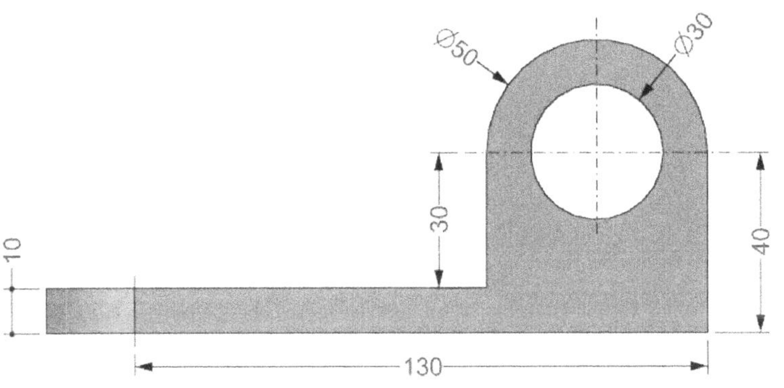

Ø50
Ø30
30
40
10
130

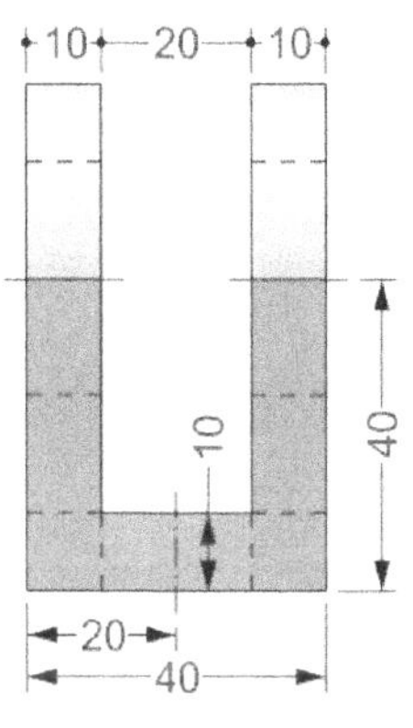

10
20
10
10
40
20
40

EX-06

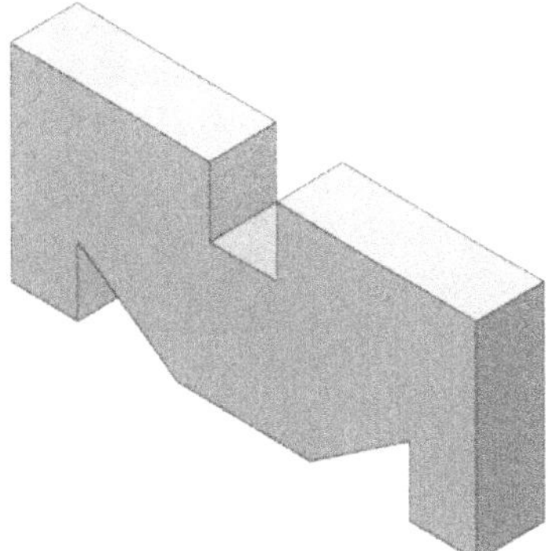

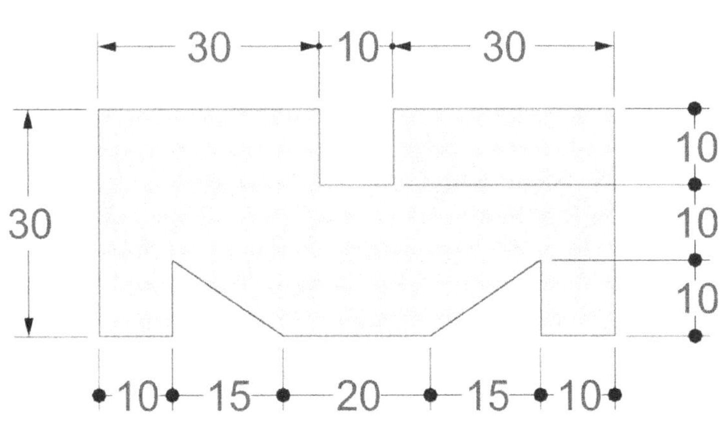

30
10
30
10
10
10
30
10
15
20
15
10

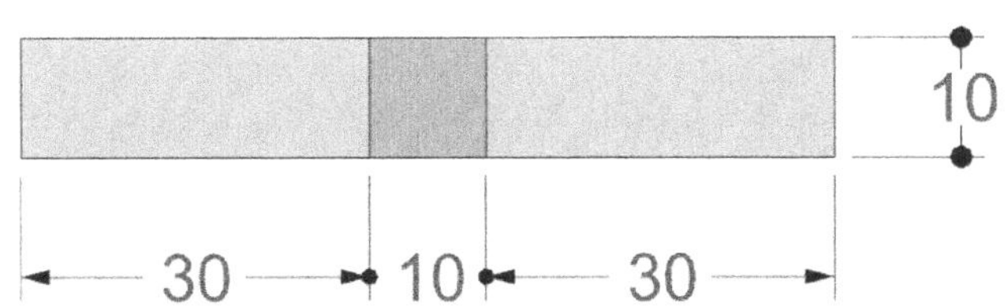

10
30
10
30

P-03

EX-07

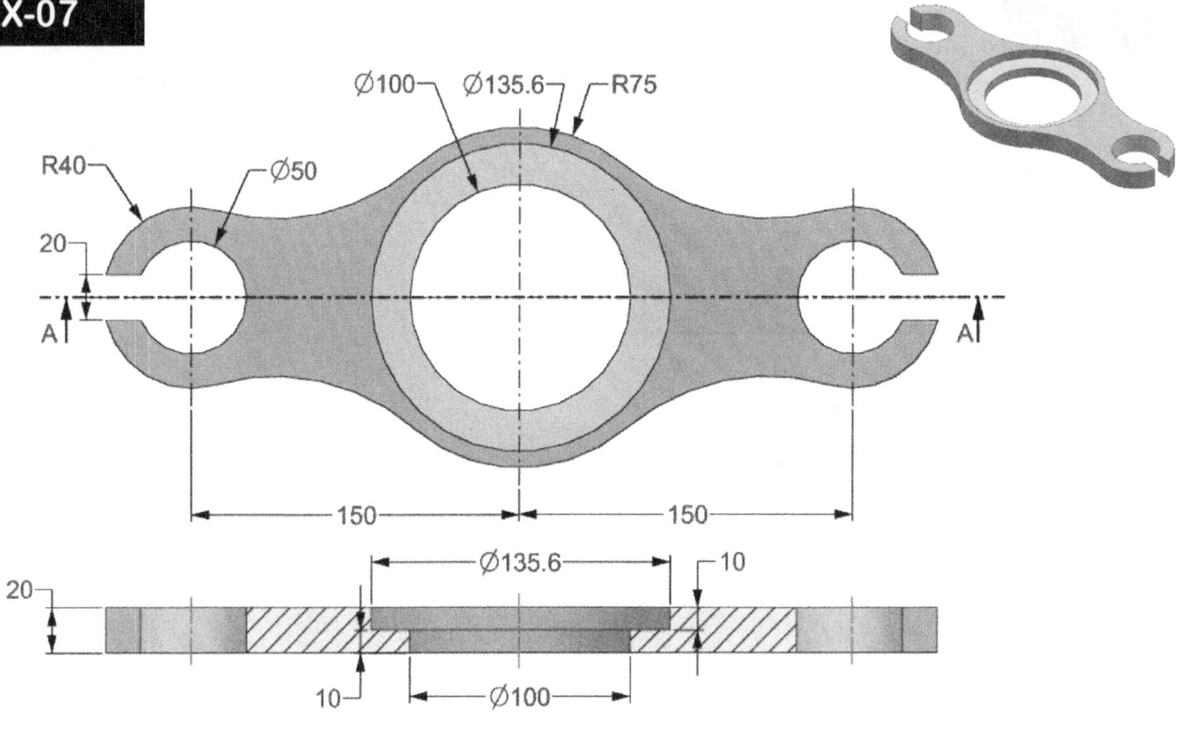

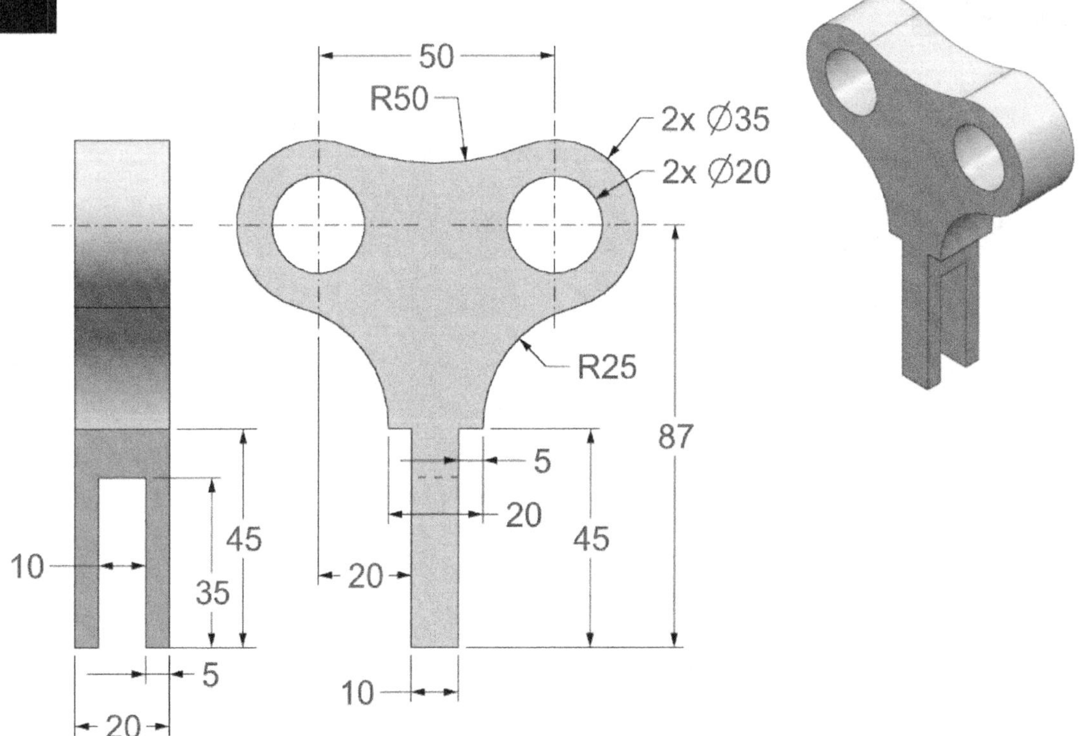
EX-07
Ø100
Ø135.6
R75
R40
Ø50
20
A
A
150
150
20
Ø135.6
10
10
Ø100
SECTION A-A
(SCALE 1:1)
EX-08
50
R50
2x Ø35
2x Ø20
R25
87
5
20
45
20
10
45
35
10
5
20
P-04

EX-08

EX-09

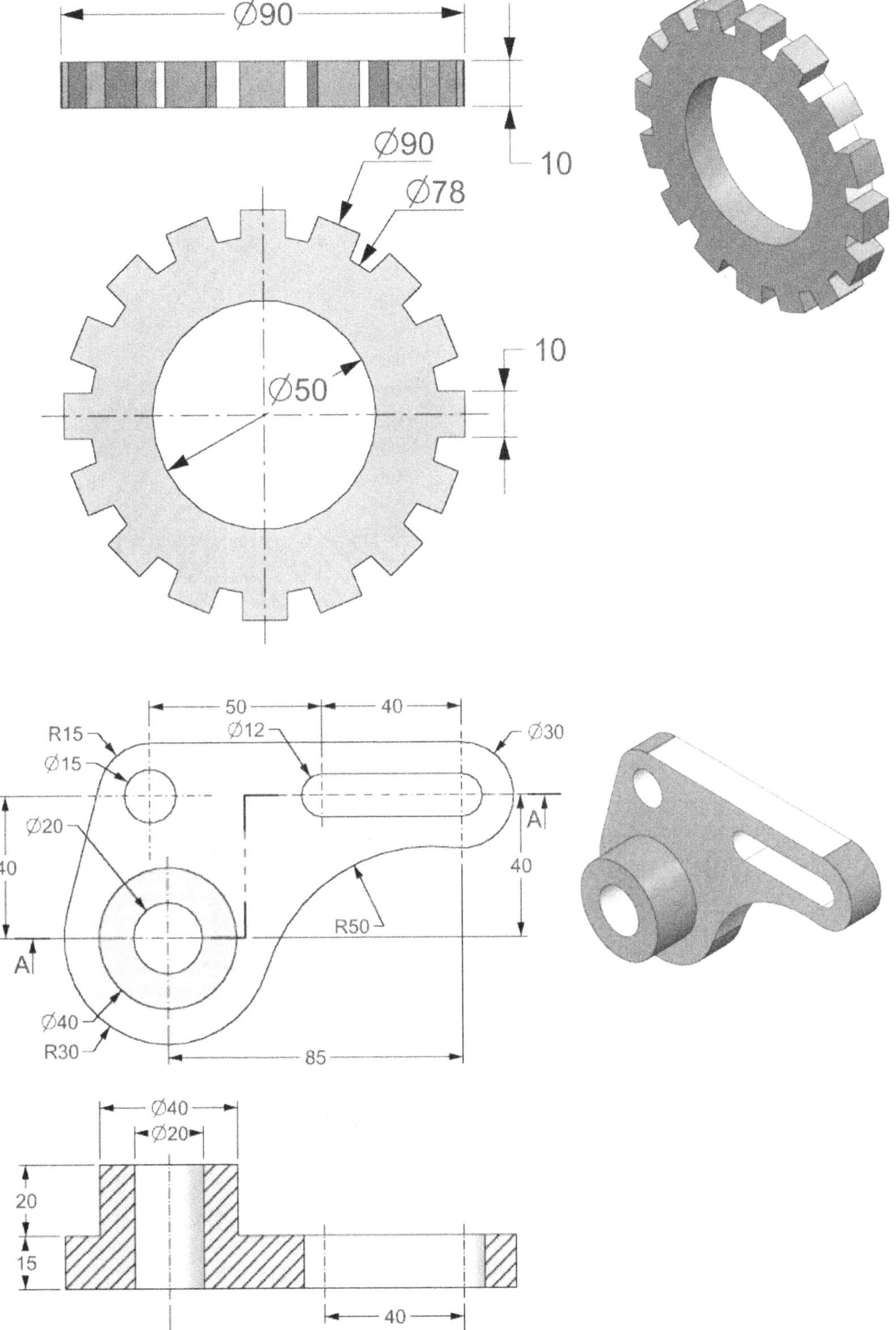

Ø90
Ø90
Ø78
Ø50
10
10
EX-10
50
40
R15
Ø15
Ø12
Ø30
Ø20
A
40
40
R50
A
R30
Ø40
85
Ø40
Ø20
20
15
40
85
SECTION A-A
(SCALE 1:1)
P-05

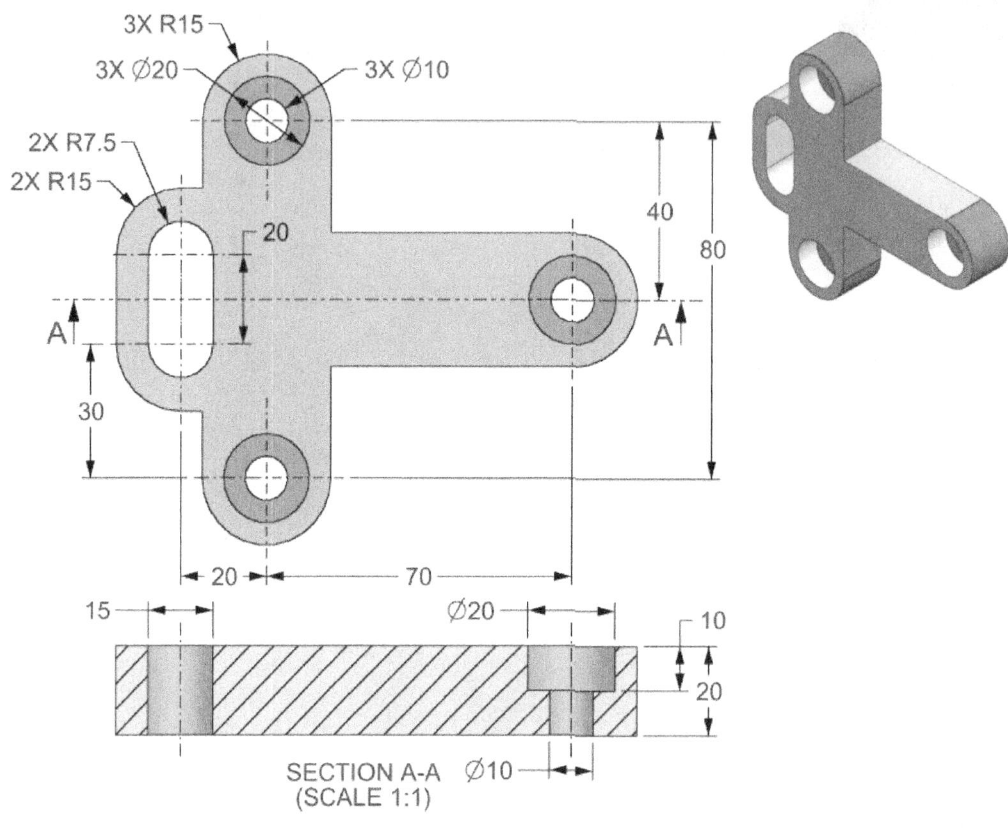

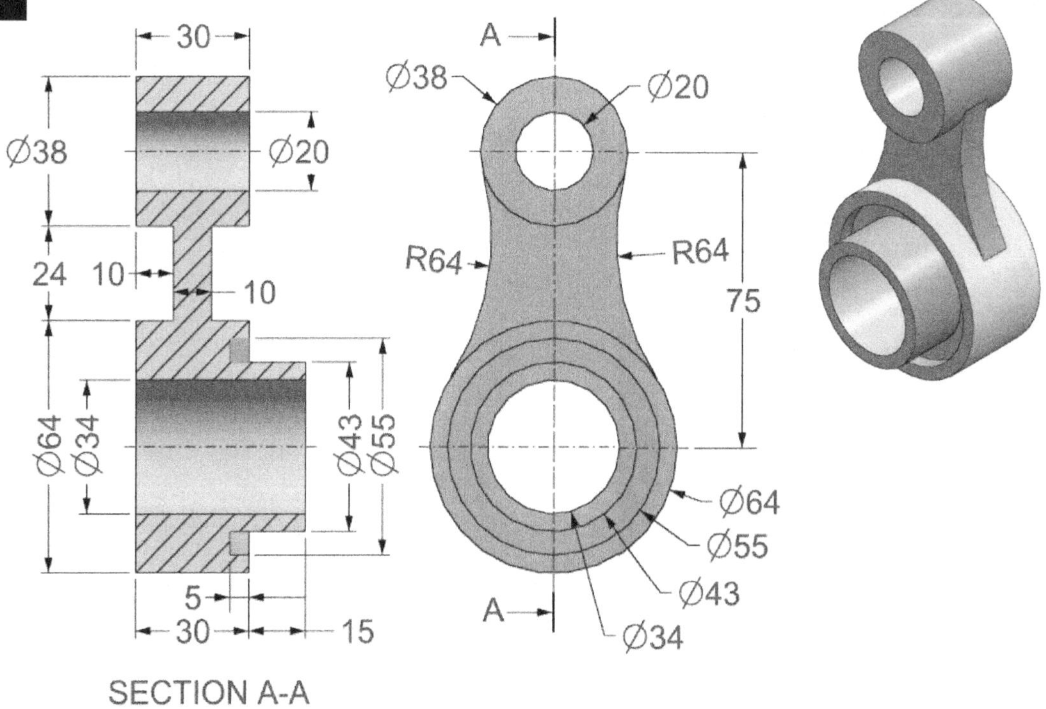

SECTION A-A
(SCALE 1:1)

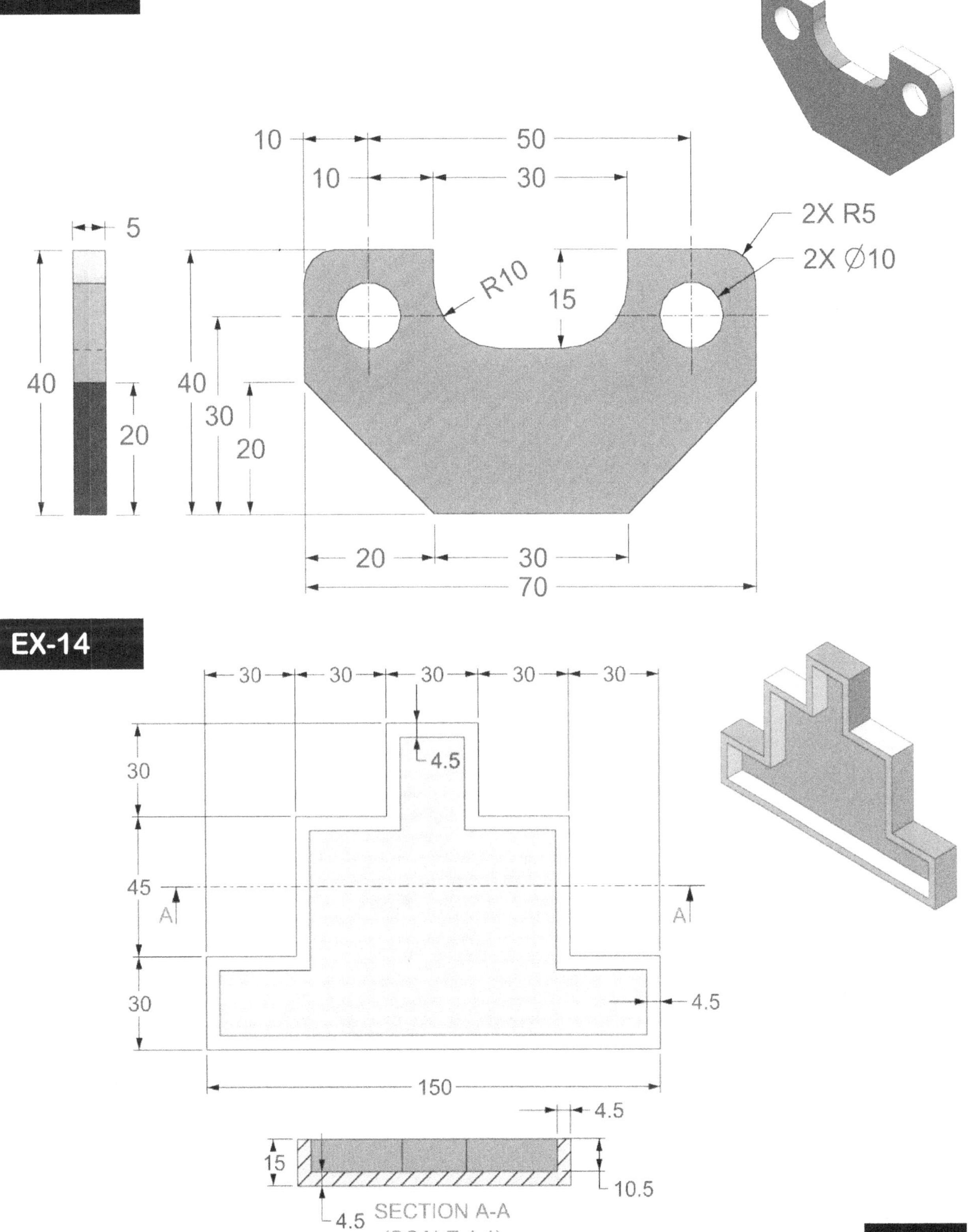

EX-13
5
10
50
10
30
2X R5
2X Ø10
R10
15
40
40
30
20
20
20
30
70
EX-14
30
30
30
30
30
30
4.5
30
45
A
A
30
4.5
150
4.5
15
10.5
4.5
SECTION A-A
(SCALE 1:1)
P-07

EX-15
4X R40
4X Ø60
4X Ø40 THRU HOLES
2X R54
10
200
100
100
50
50
40
10
2X Ø40
A
100
280
180
200
100
100
50
50
20
100
280
Ø60
Ø40
5
Ø40
25
25
20
280
SECTION A-A
EX-16
Ø50
Ø34
13.8
R32
20
40
90
26
A
A
164.9
Ø34
32
10
(SCALE 1:1) SECTION A-A
P-08

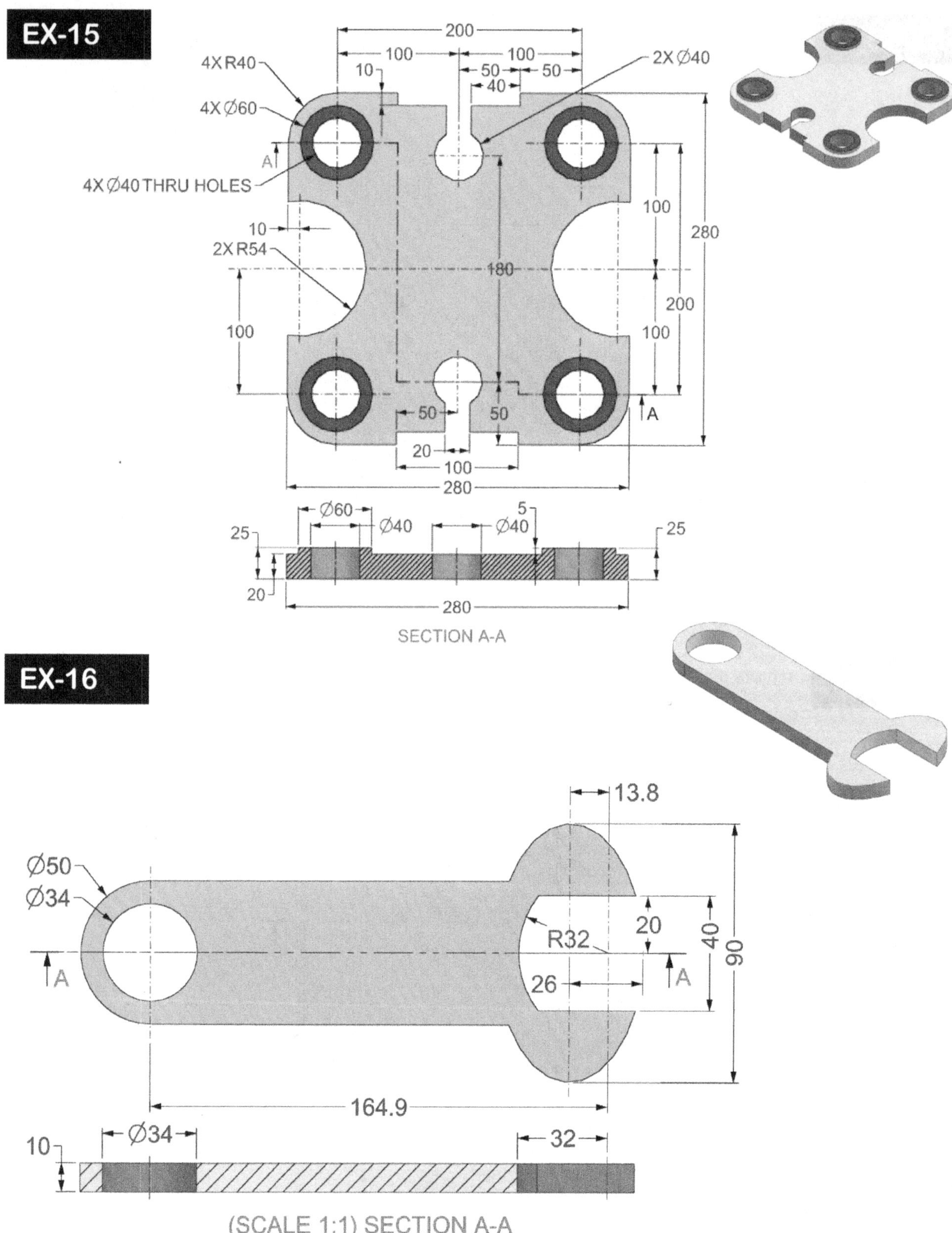

EX-17

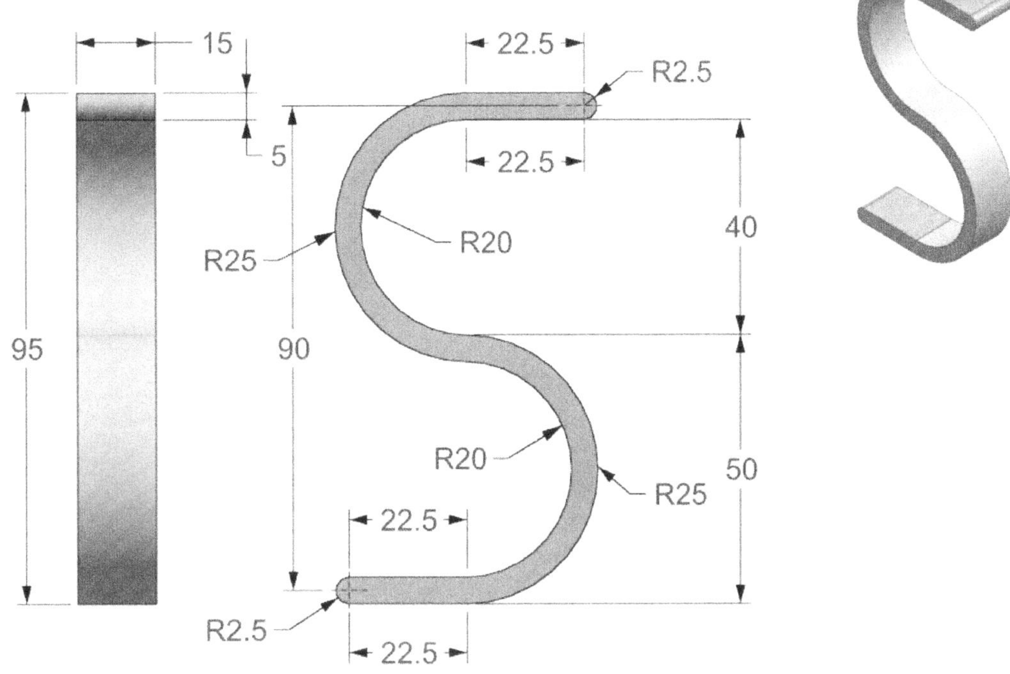

15
22.5
R2.5
5
22.5
R20
R25
40
95
90
R20
50
R25
22.5
R2.5
22.5

EX-18

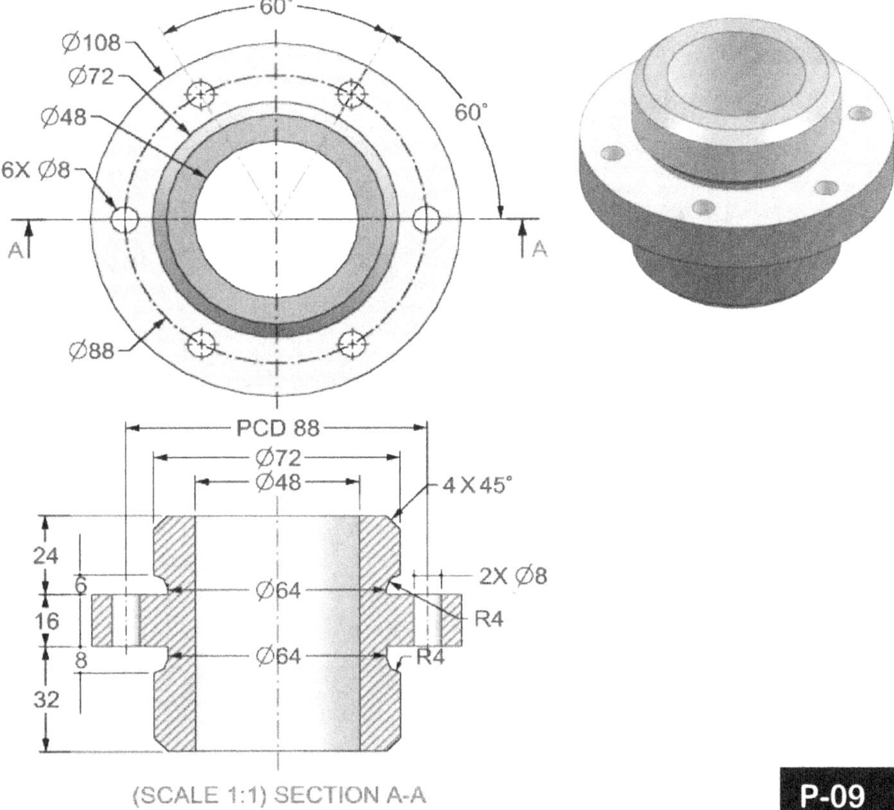

60°
Ø108
Ø72
Ø48
6X Ø8
60°
A
A
Ø88
PCD 88
Ø72
Ø48
4 X 45°
24
6
Ø64
2X Ø8
16
R4
8
Ø64
R4
32
(SCALE 1:1) SECTION A-A

P-09

EX-19

5
A
6X Ø8 THRU HOLES ON PCD 60
6X R10
PCD Ø60
Ø8
80
Ø32
60
72
Ø8
R20
Ø32
6X R6
10
A
SECTION A-A

EX-20

216
12.3
126
12.7
R41
Ø60
63
4X R44.8
4X R32.5
214
126
63
A
146
A
4X R12
Ø60
25
146
SECTION A-A

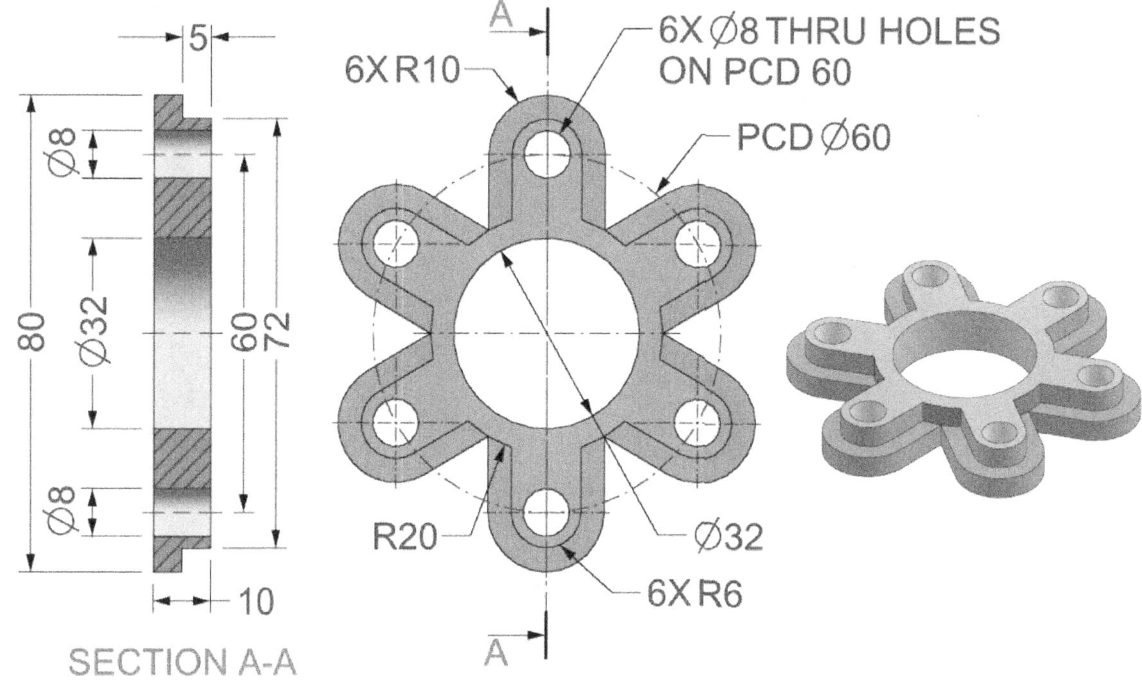

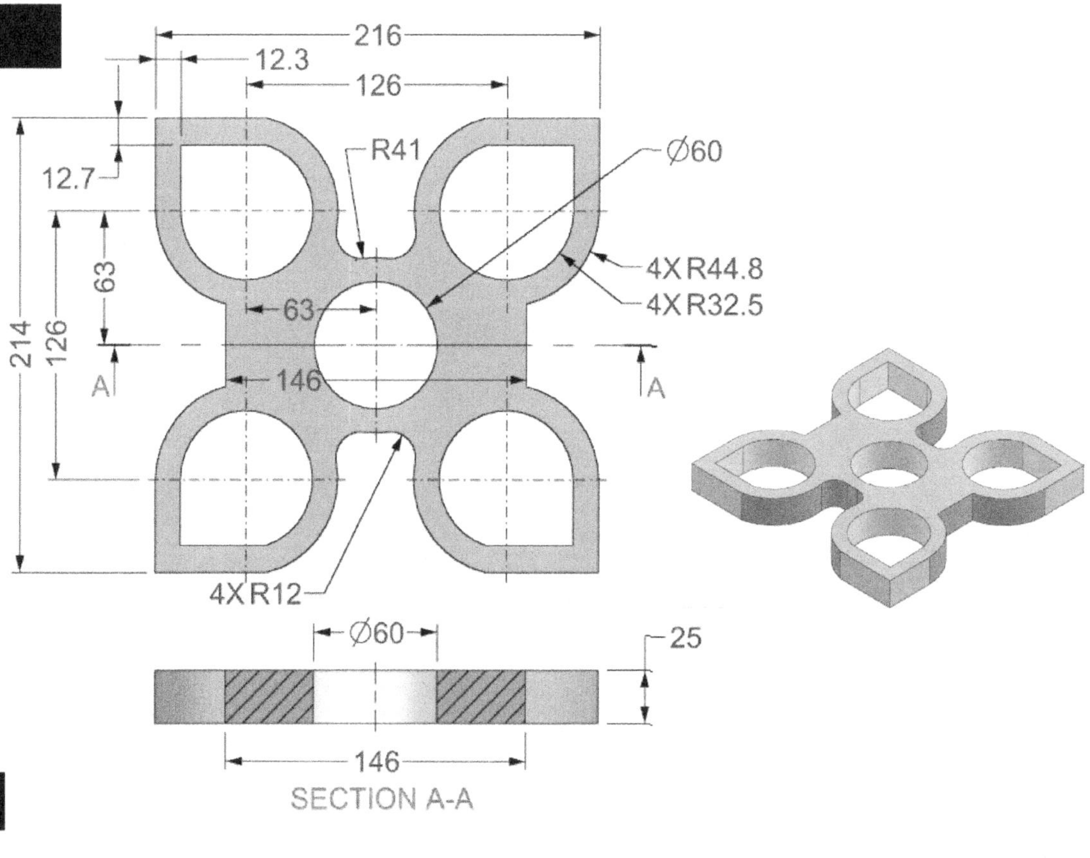

EX-21
150 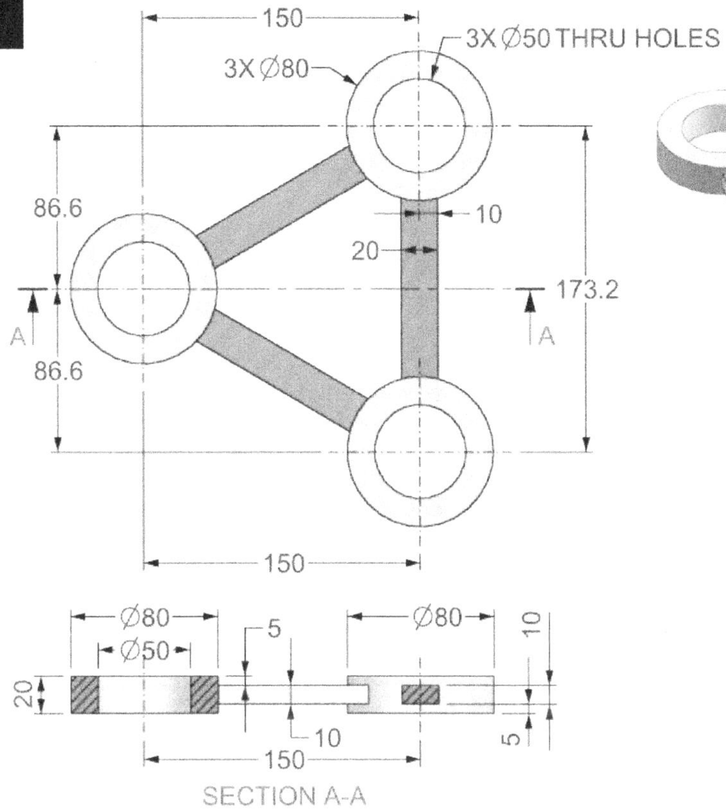
3X Ø80
3X Ø50 THRU HOLES
86.6
10
20
173.2
A
A
86.6
150
Ø80
Ø50
5
Ø80
10
20
10
5
150
SECTION A-A

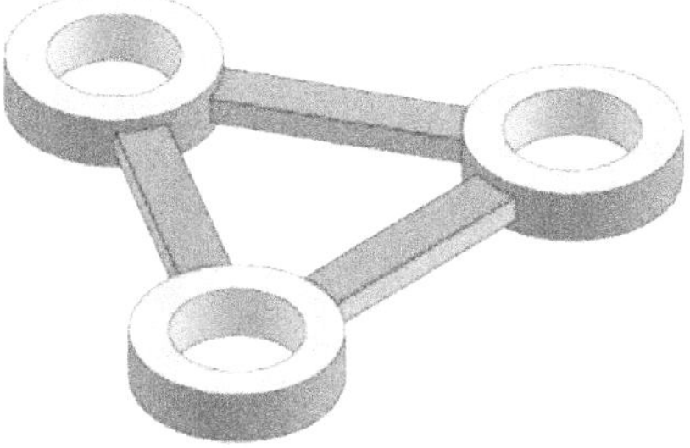

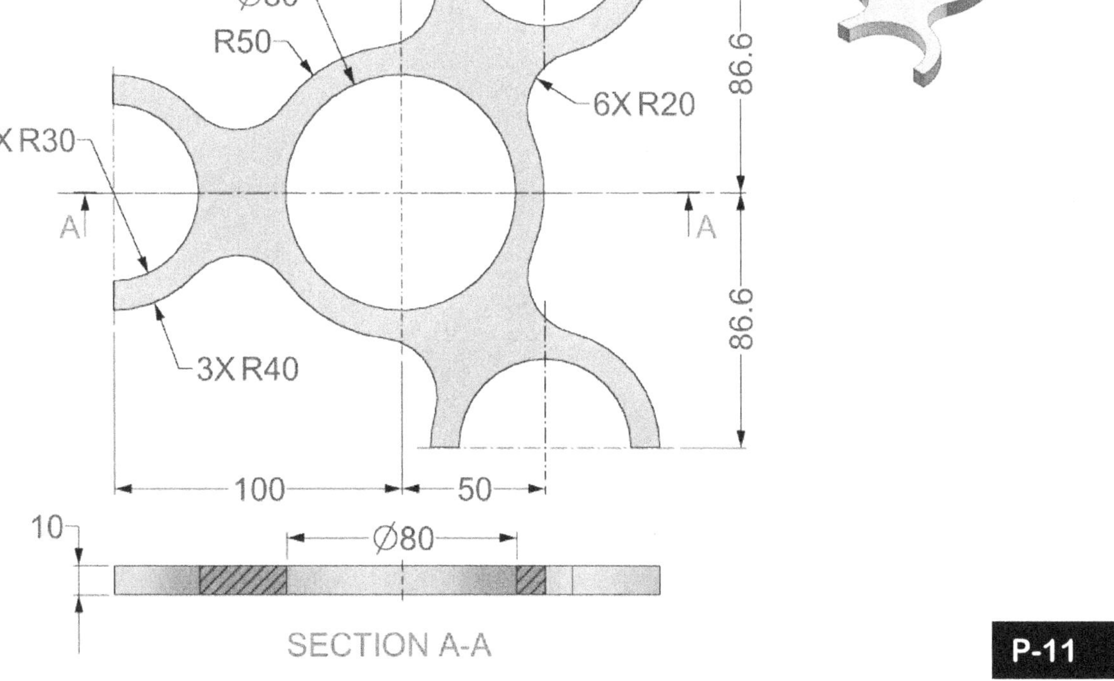
EX-22
50
Ø80
R50
86.6
6X R20
3X R30
A
A
3X R40
86.6
100
50
10
Ø80
SECTION A-A
P-11

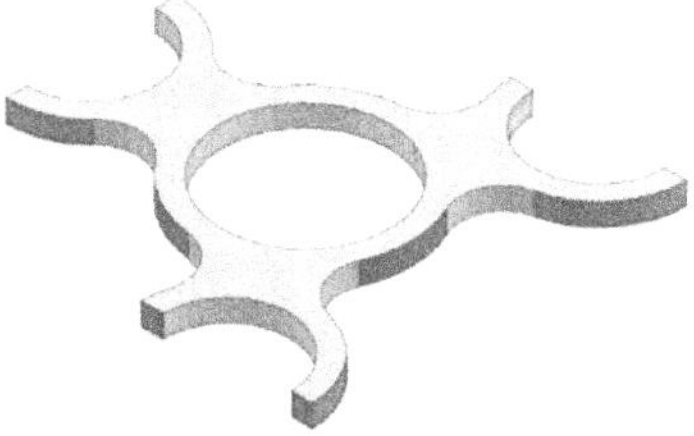

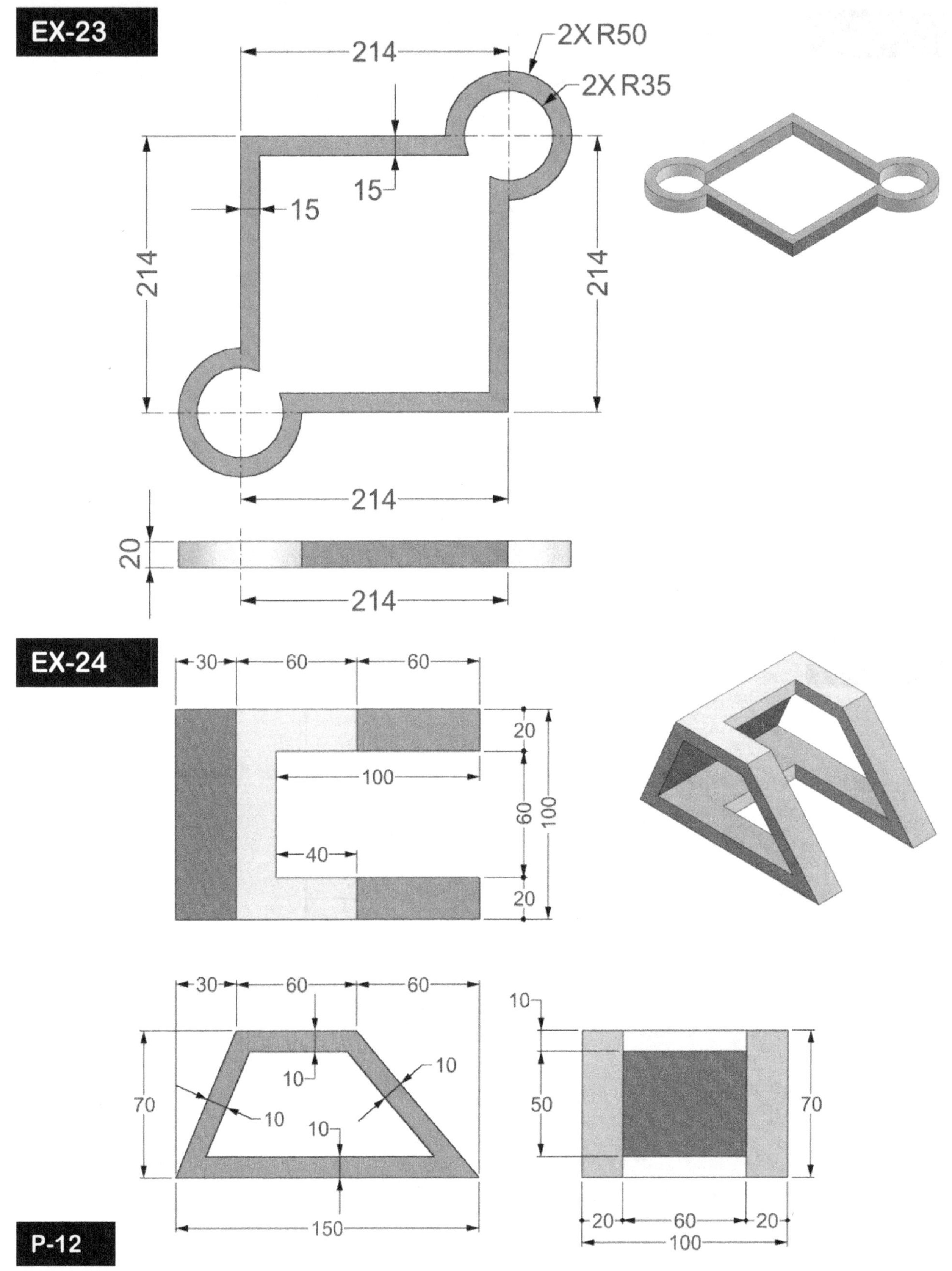

EX-23
214
2X R50
2X R35
15
15
214
214
214
214
20
EX-24
30
60
60
20
100
60
100
40
20
P-12
30
60
60
10
10
10
70
10
10
150
10
50
70
20
60
20
100

EX-25
2X Ø50
2X Ø80
Ø150
Ø120
R100
R100
Ø100
150
150
Ø80
Ø50
Ø120
Ø100
20
50
20
70
40
150
150
SECTION A-A
(SCALE 1:1)
A
A

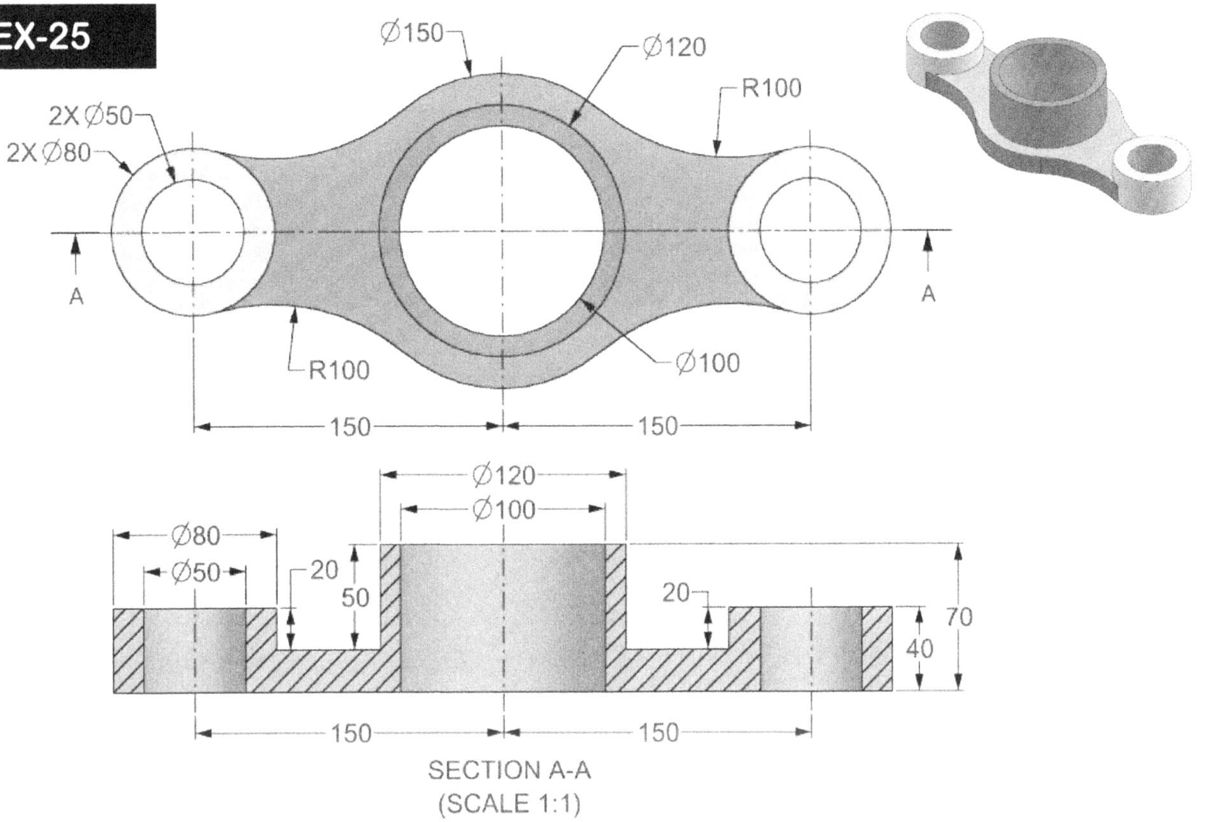

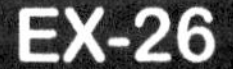

EX-26
Ø24
Ø44
Ø36
A
A
Ø44
Ø36
Ø32
Ø24
2X45°
4
3
12
8
12
36
SECTION A-A
(SCALE 1:1)

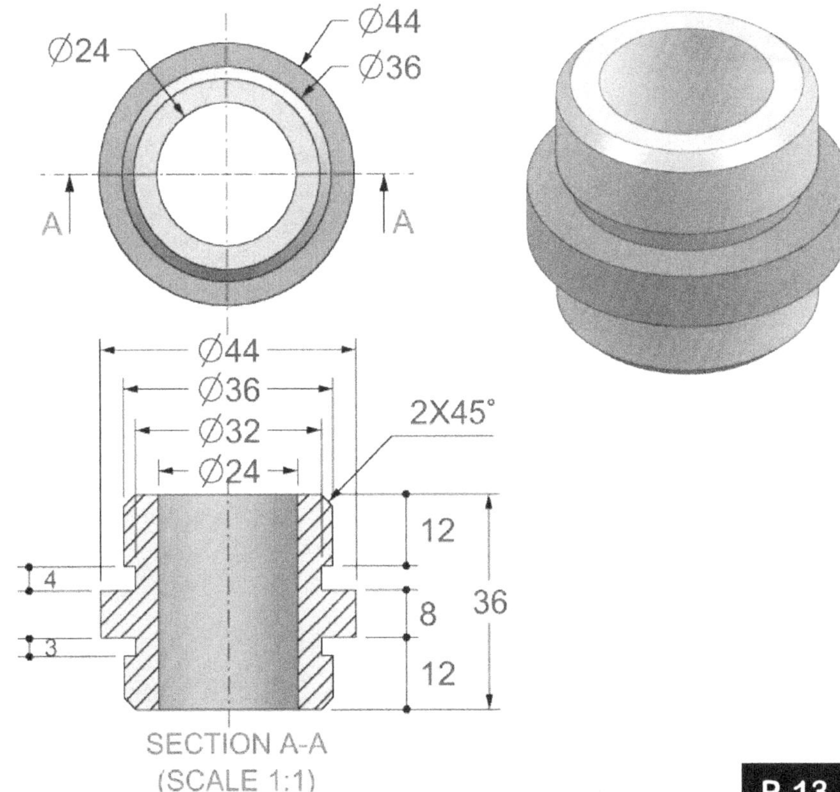

P-13

EX-27 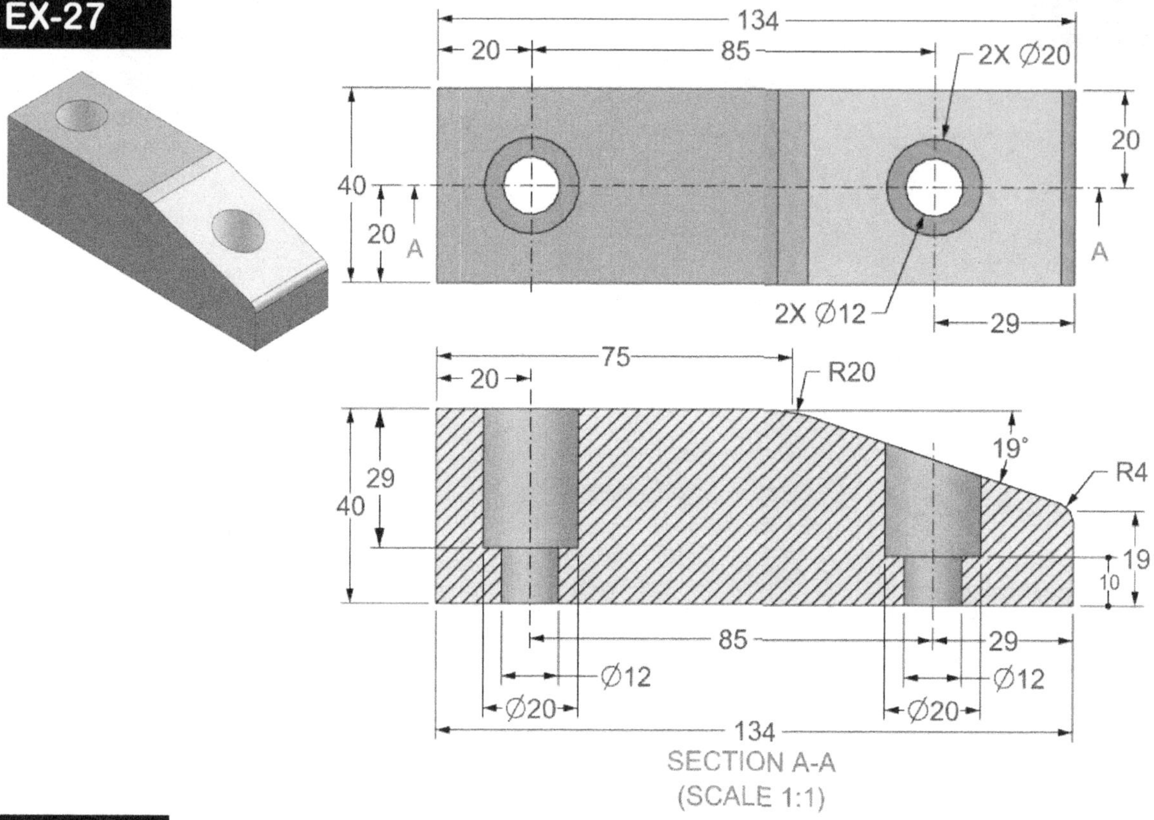
134
20
85
2X Ø20
40
20
20
A
A
2X Ø12
29
75
20
R20
29
19°
40
R4
19
10
85
29
Ø12
Ø12
Ø20
Ø20
134
SECTION A-A
(SCALE 1:1)

EX-28
Ø120
18X Ø10
PCD Ø70
60°
A
A
PCD Ø40
Ø20
PCD Ø100
10
Ø120
SECTION A-A

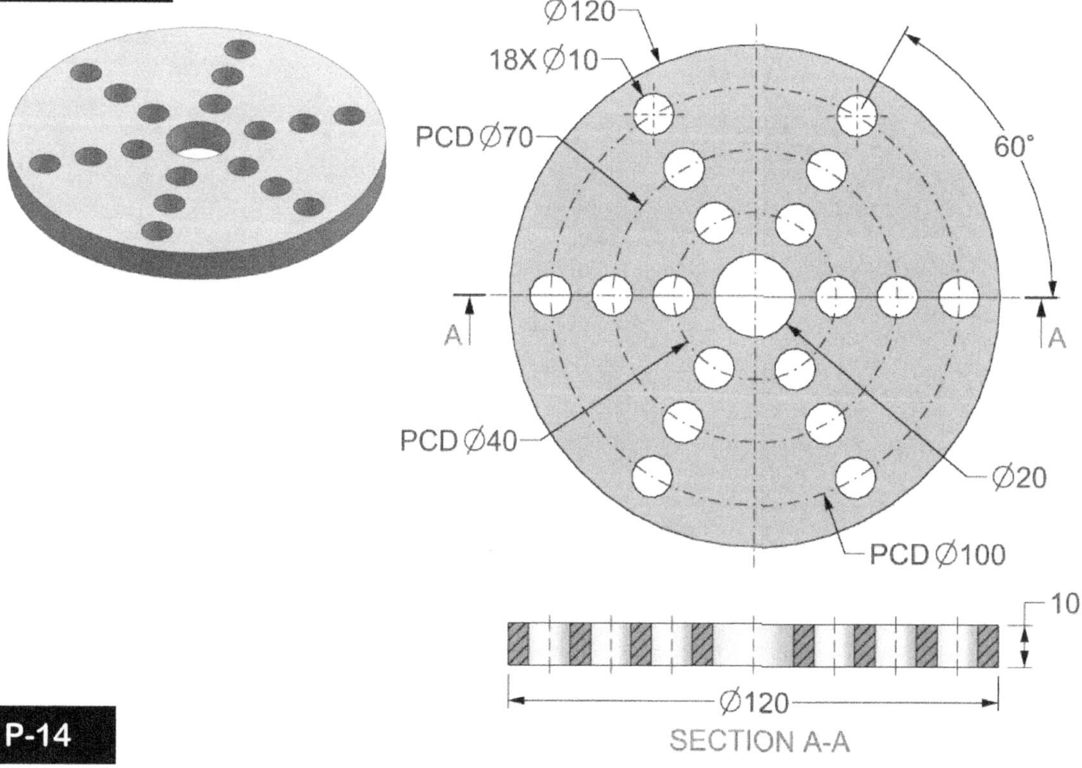

P-14

EX-29

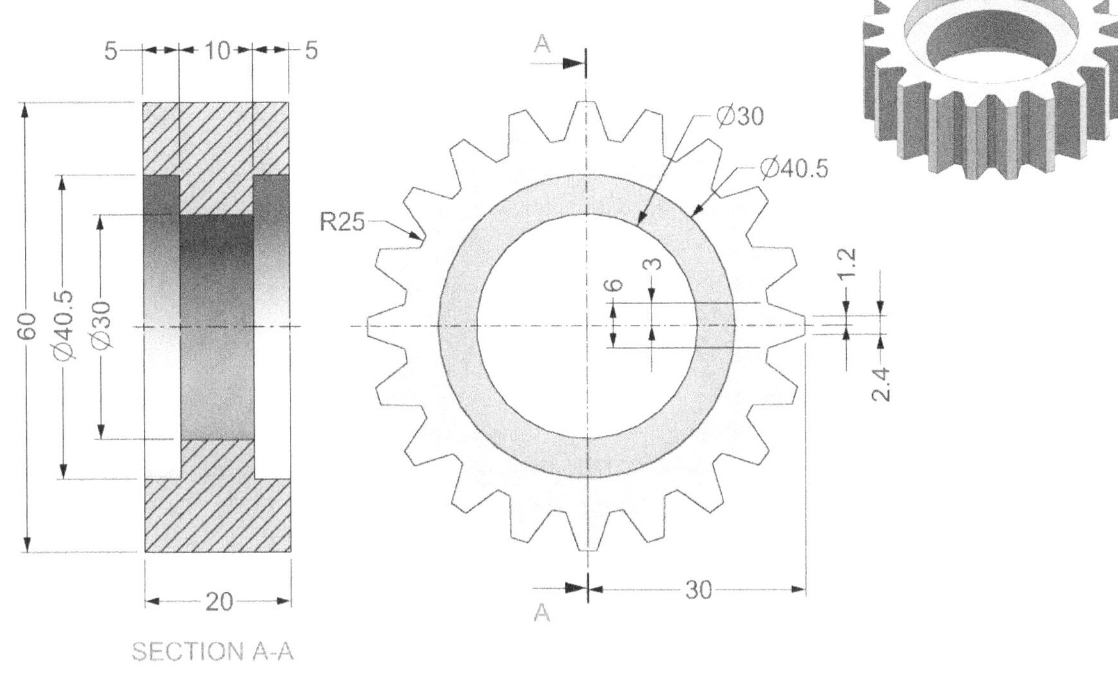
5
10
5
A
Ø30
Ø40.5
R25
60
Ø40.5
Ø30
6
3
1.2
2.4
20
30
A
SECTION A-A

EX-30
90
10
70
10
10
50
10
10
30
10
5
Ø5
10
10
Ø10
20
40
10
5
10
10
Ø50
10
Ø10
10
25
40
20
Ø5
10
25
40
10
10
10
5
10
20
10
10
30
10
10
5
10
10
30
10
10
10
70
10
90
10
20
10
5
10
10
5
5
40

P-15

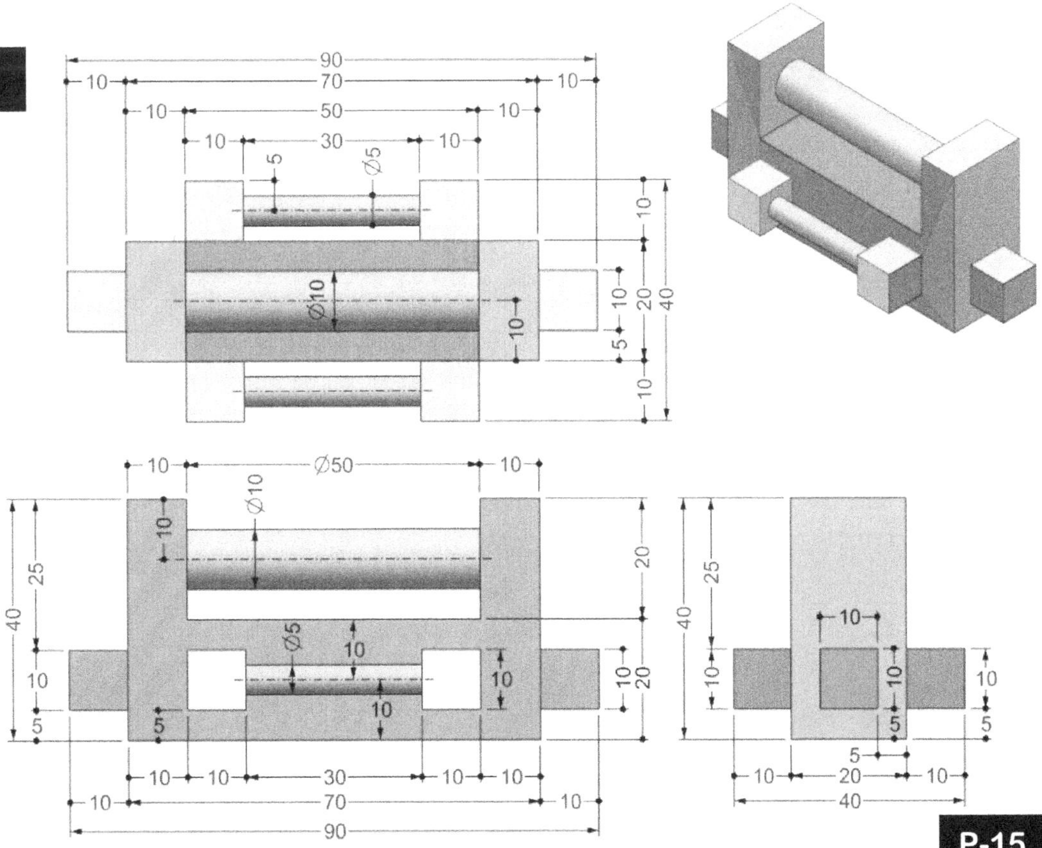

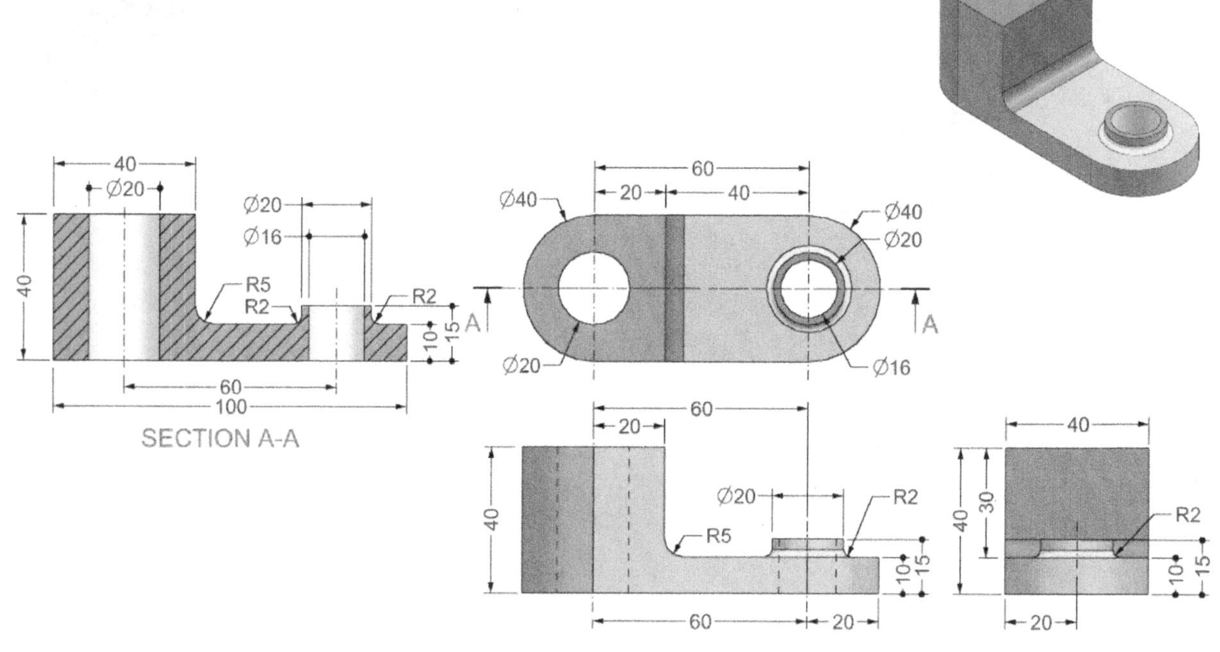

EX-31

40
Ø20
Ø20
Ø16
R5
R2
R2
40
10
15
60
100
SECTION A-A
60
20
40
Ø40
Ø40
Ø20
A
A
Ø20
Ø16
60
20
Ø20
R2
R5
40
10
15
60
20
40
30
R2
40
10
15
20

EX-32

3X Ø30
3X Ø60
40
100
70
100
100
Ø30
Ø40
Ø60
30
29.8
R5
10
Ø60
30
30
20
20
30
100
100
70
69.8
49.8
30
Ø60
Ø60
20
30
100

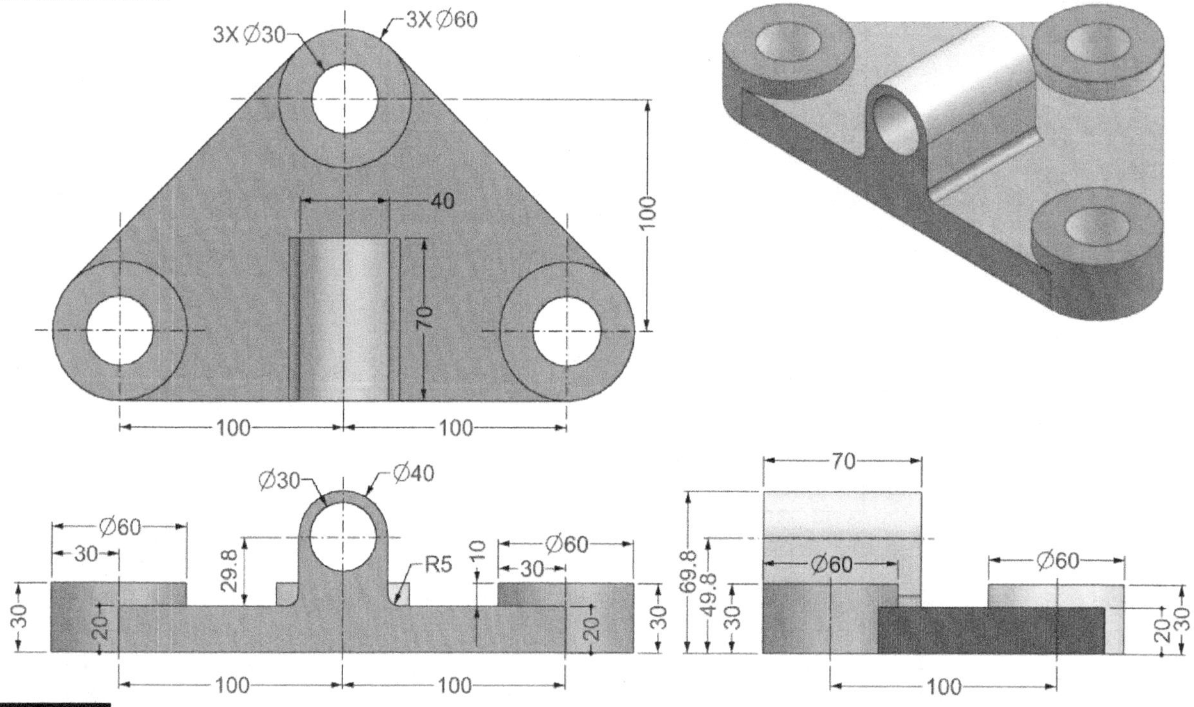

EX-33

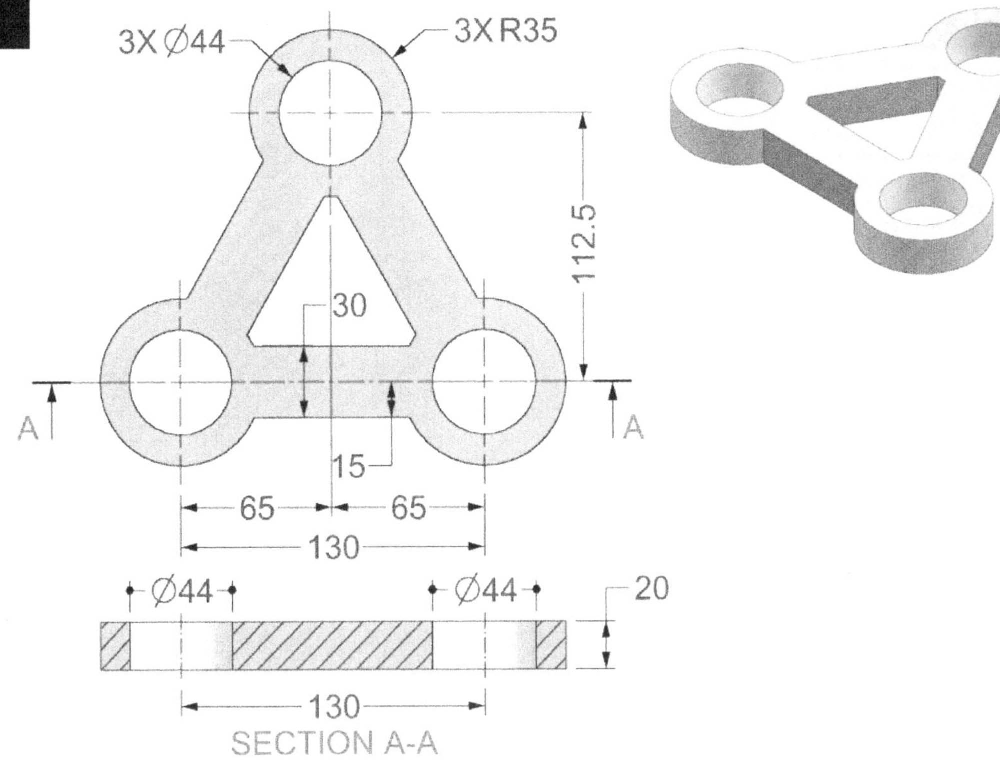
3X Ø44
3X R35
112.5
30
15
A
A
65
65
130
Ø44
Ø44
20
130
SECTION A-A

EX-34

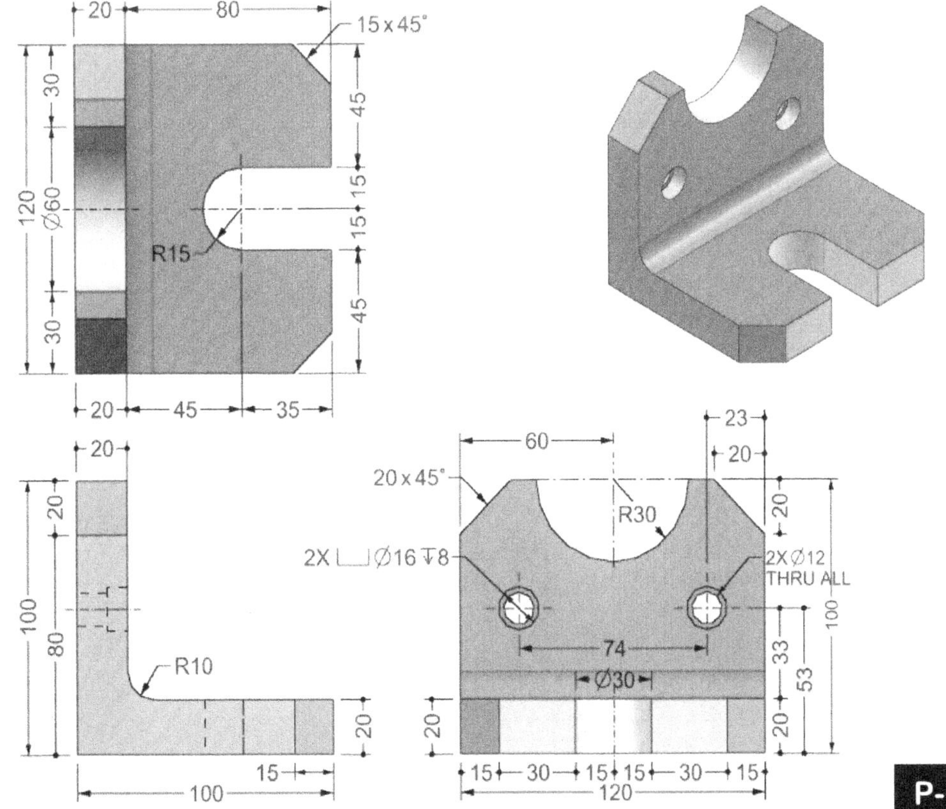
20
80
15 x 45°
30
45
120
Ø60
15 15
15
R15
45
30
20
45
35
20
20
100
80
R10
20
15
100
60
20 x 45°
R30
23
20
20
2X ⌴ Ø16 ↧8
2X Ø12
THRU ALL
74
100
Ø30
33
53
20
15 30 15 15 30 15
120

P-17

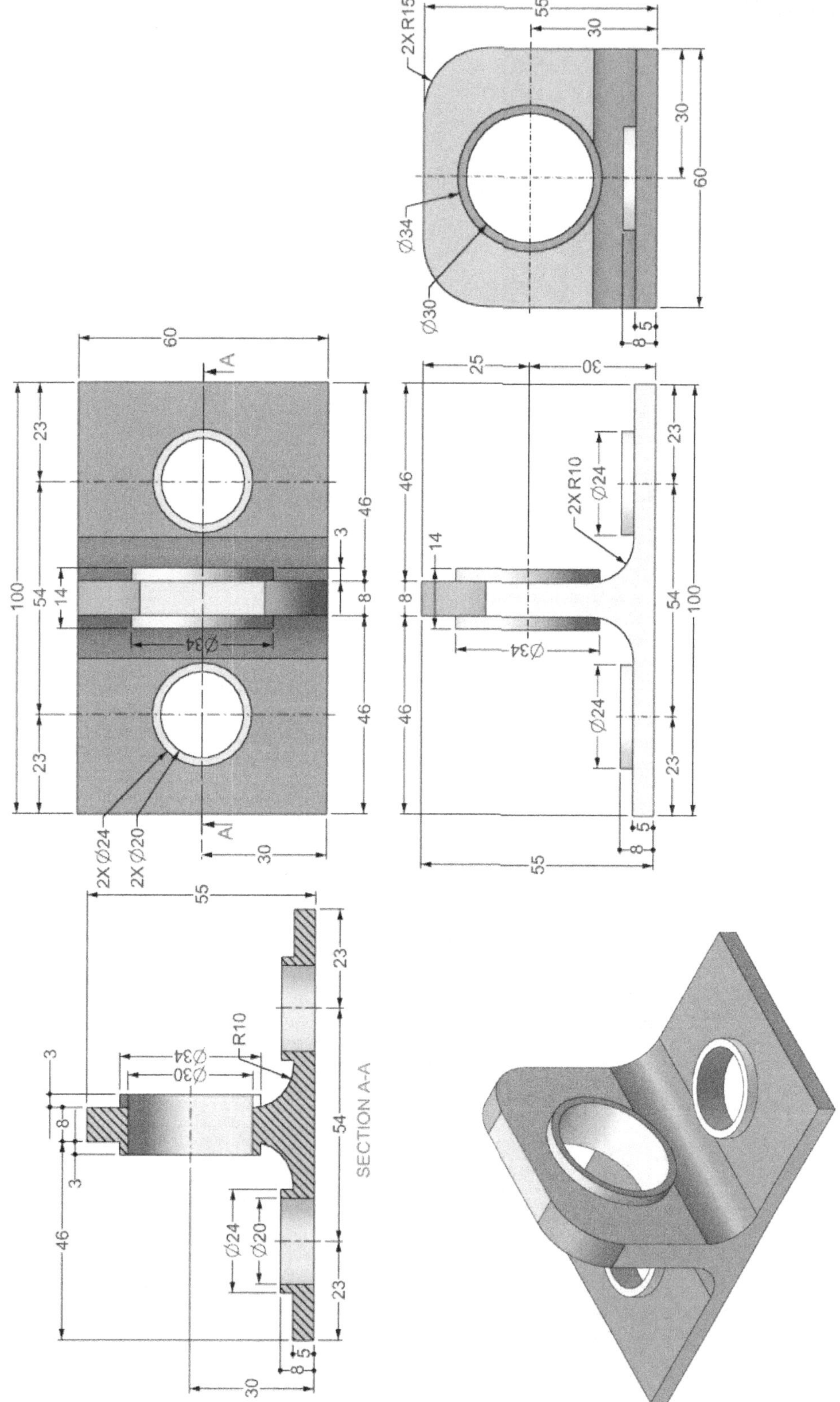

2X R15
Ø34
Ø30
55
30
30
60
8
5
60
A
23
46
3
100
54
14
8
Ø34
46
23
30
2X Ø24
2X Ø20
A
25
30
46
14
8
2X R10
Ø24
23
54
100
Ø24
46
23
8
5
55
55
3
Ø34
Ø30
R10
23
8
54
3
46
Ø24
Ø20
23
8
5
30
SECTION A-A

EX-36

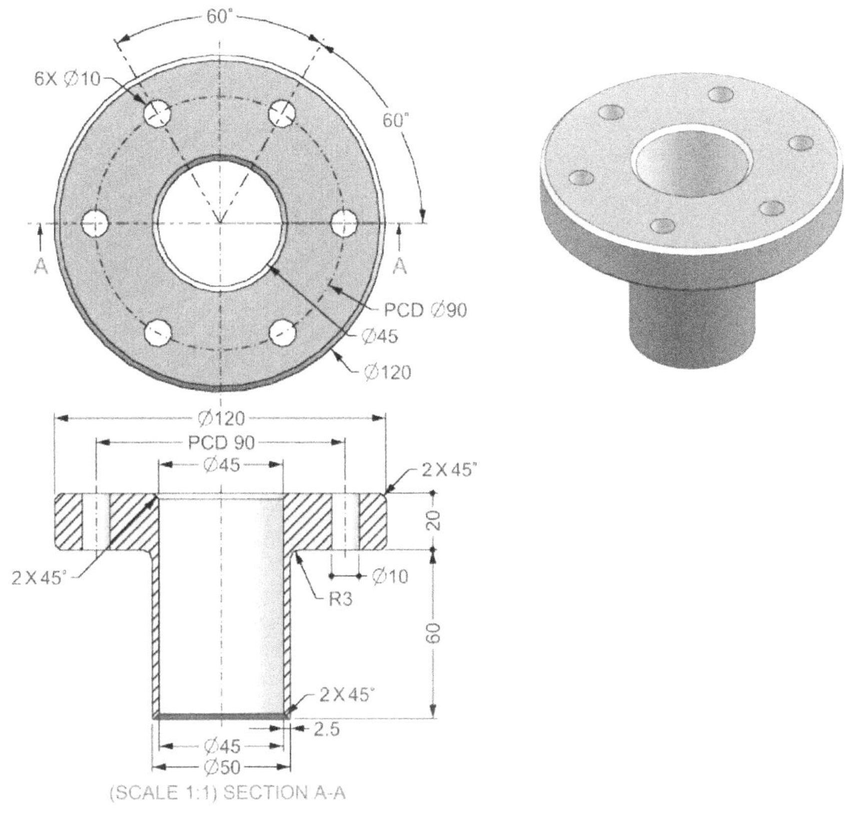

60°
6X Ø10
60°
PCD Ø90
Ø45
Ø120
Ø120
PCD 90
Ø45
2 X 45°
20
2 X 45°
Ø10
R3
60
2 X 45°
2.5
Ø45
Ø50
(SCALE 1:1) SECTION A-A
A
A

EX-37 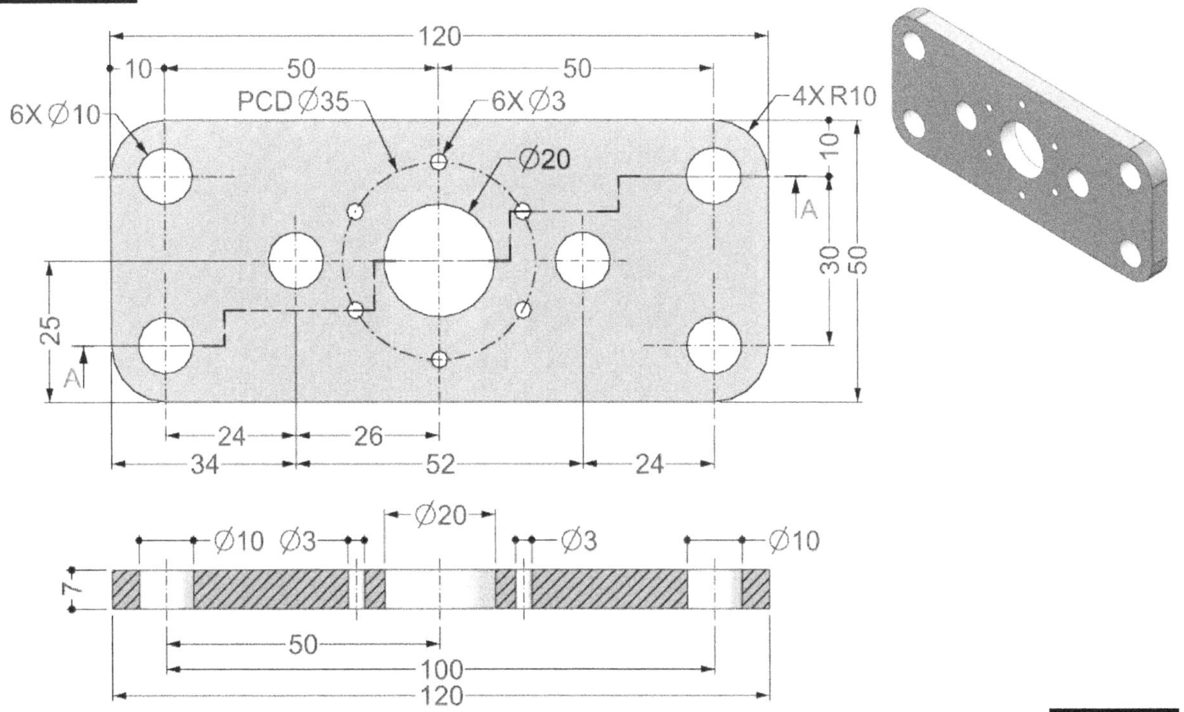
120
10
50
50
6X Ø10
PCD Ø35
6X Ø3
4X R10
Ø20
10
A
30
50
25
A
24
26
34
52
24
Ø20
Ø10 Ø3
Ø3
Ø10
7
50
100
120
SECTION A-A

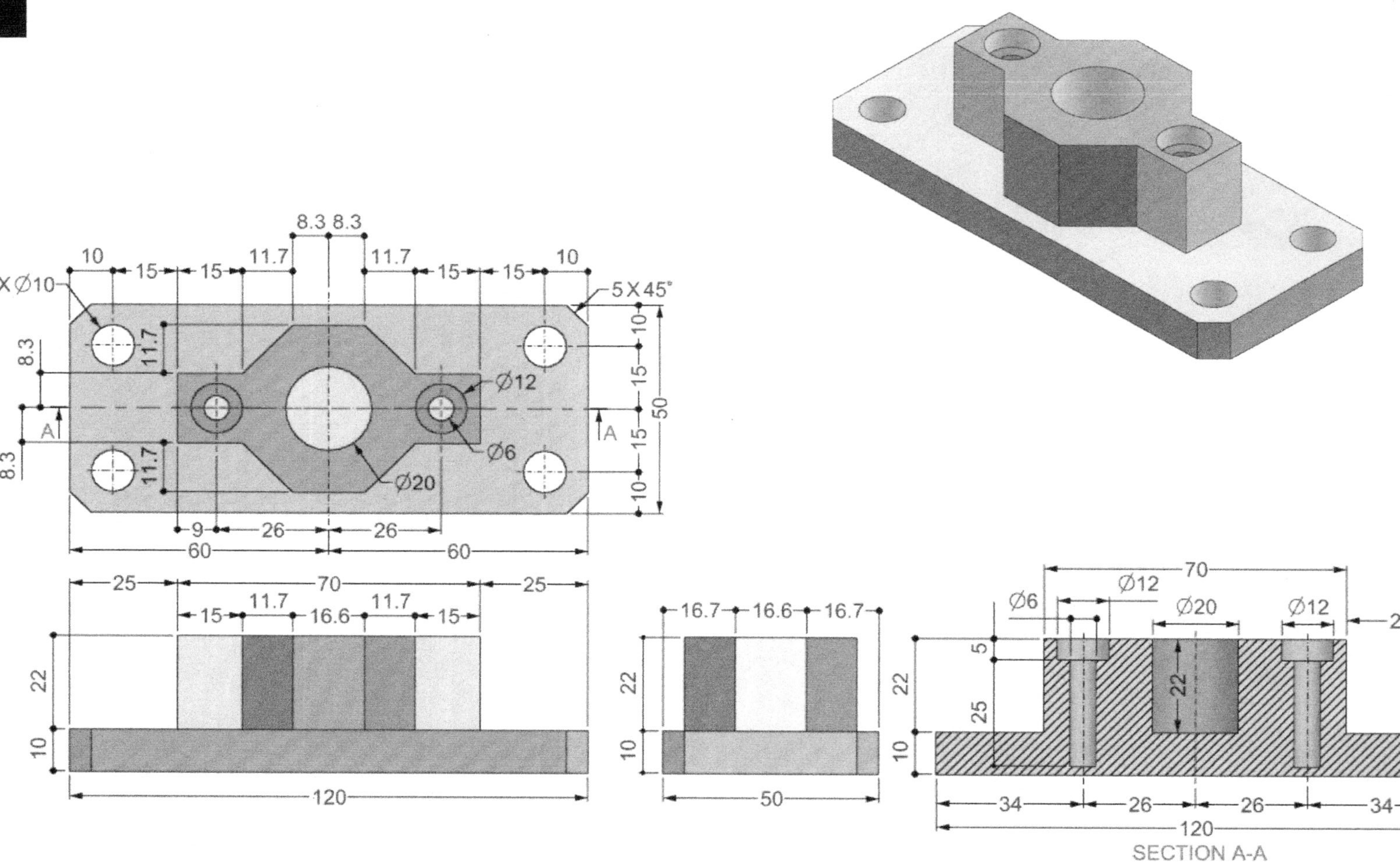

P-20
EX-38
8.3 8.3
10
15
15
11.7
11.7
15
15
10
4X Ø10
5 X 45°
11.7
8.3
10
15
50
15
10
8.3
A
A
Ø12
Ø6
Ø20
9
26
26
60
60
25
70
25
15
11.7
16.6
11.7
15
22
10
120
16.7
16.6
16.7
22
10
50
Ø6
Ø12
Ø20
Ø12
70
5
25
22
22
25
34
26
26
34
120
SECTION A-A

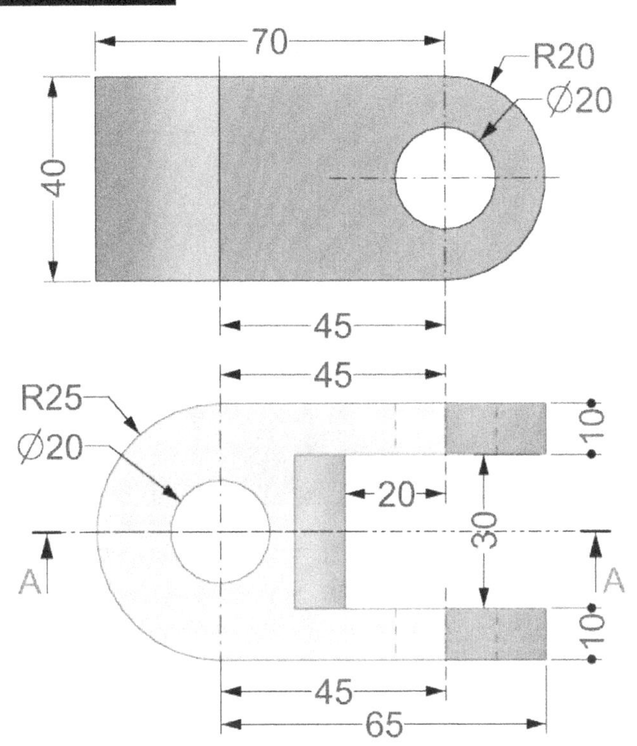
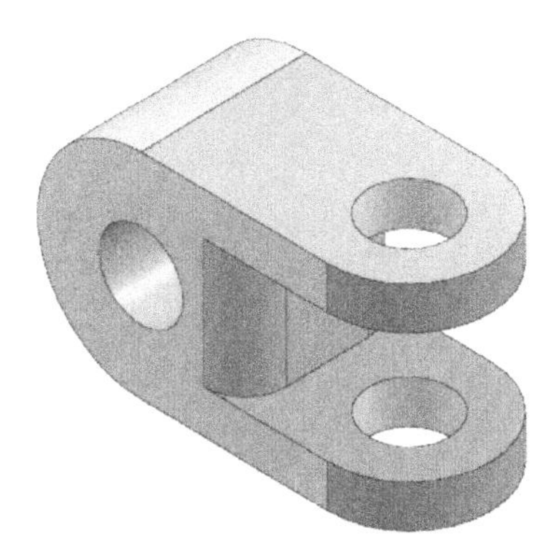
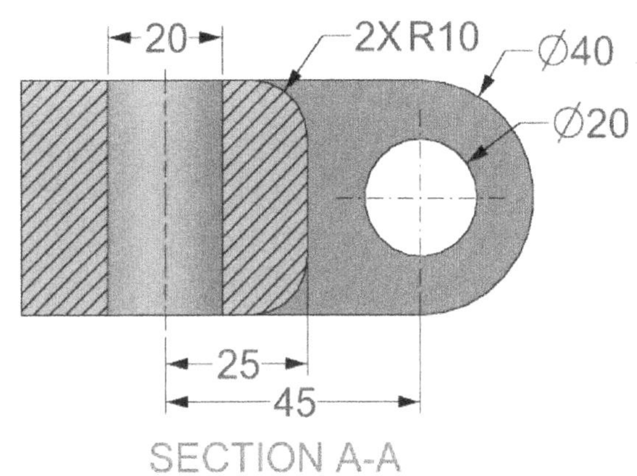

SECTION A-A

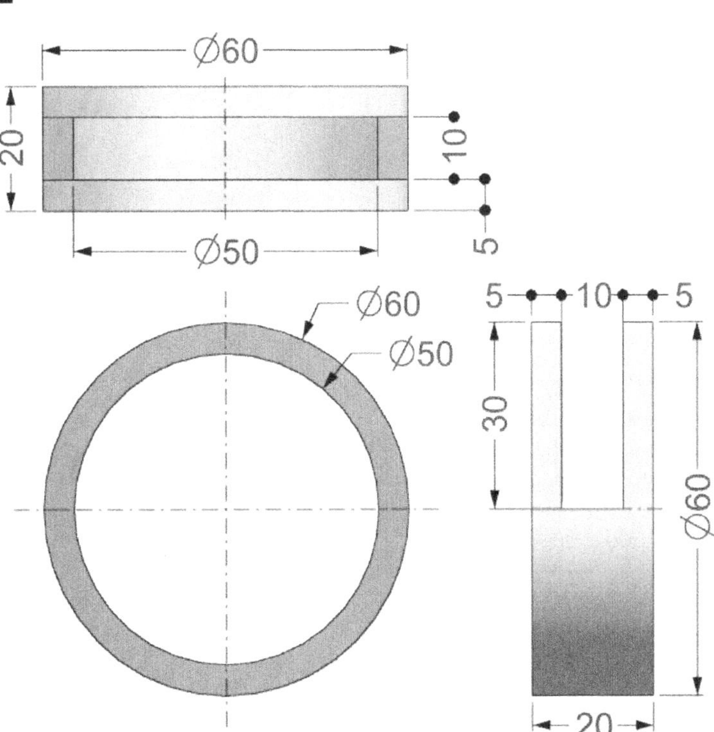
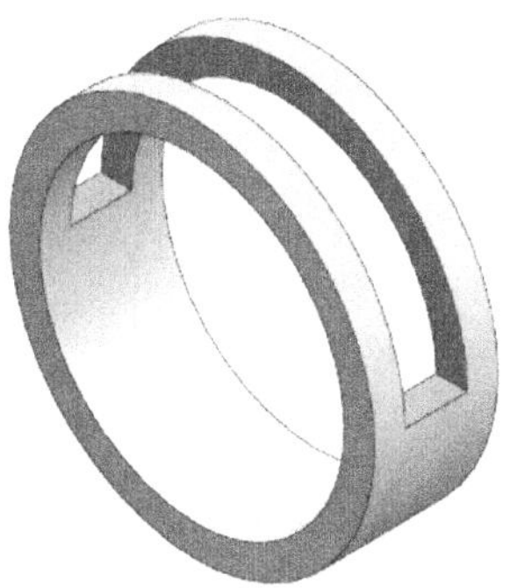

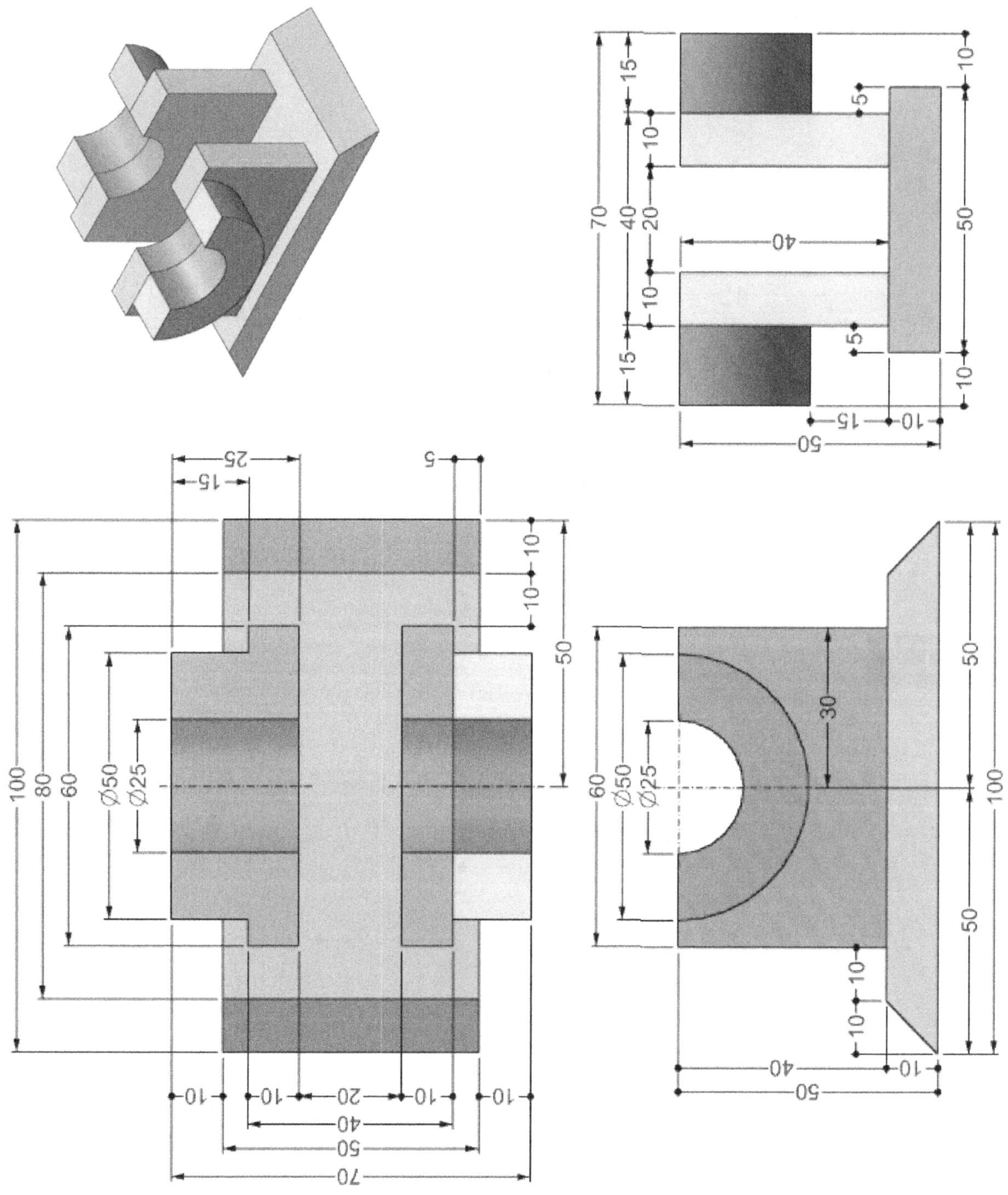

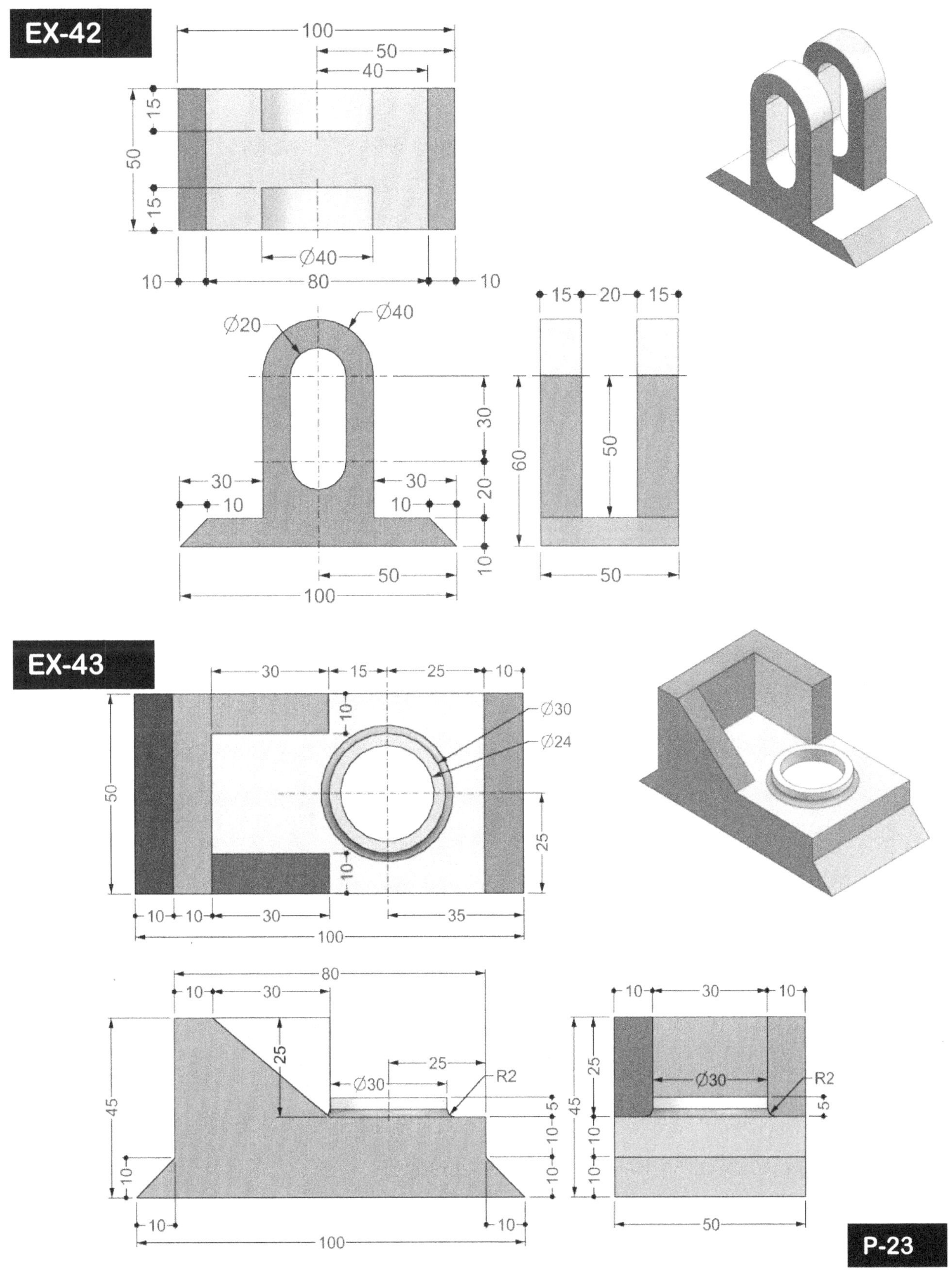

EX-42
100
50
40
15
50
15
Ø40
80
10
10
Ø20
Ø40
30
30
30
10
10
20
50
100
60
15
20
15
50
10
50
EX-43
30
15
25
10
10
Ø30
Ø24
50
25
10
10
10
30
35
100
80
10
30
25
25
Ø30
R2
45
10
10
10
10
10
100
10
30
10
25
45
Ø30
R2
10
10
5
50
P-23

EX-44

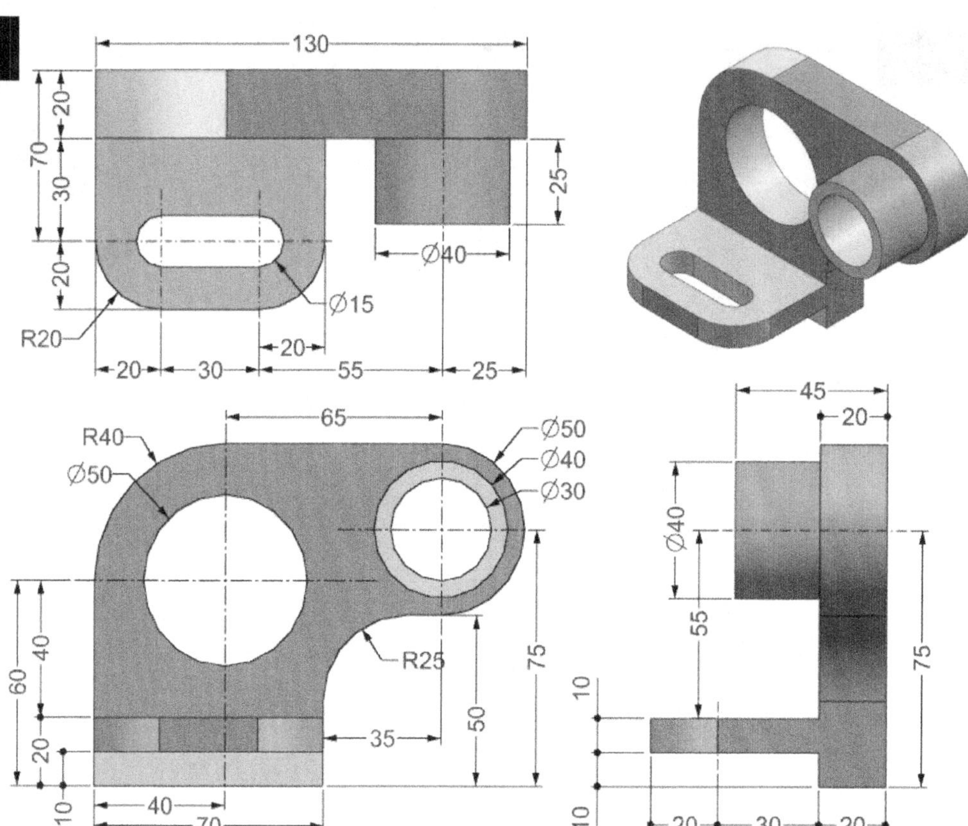
130
20
70
30
20
25
Ø40
Ø15
R20
20
30
20
55
25
R40
Ø50
Ø50
Ø40
Ø30
65
R25
60
40
20
10
75
50
35
40
70
45
20
Ø40
55
75
10
10
20
30
20

EX-45

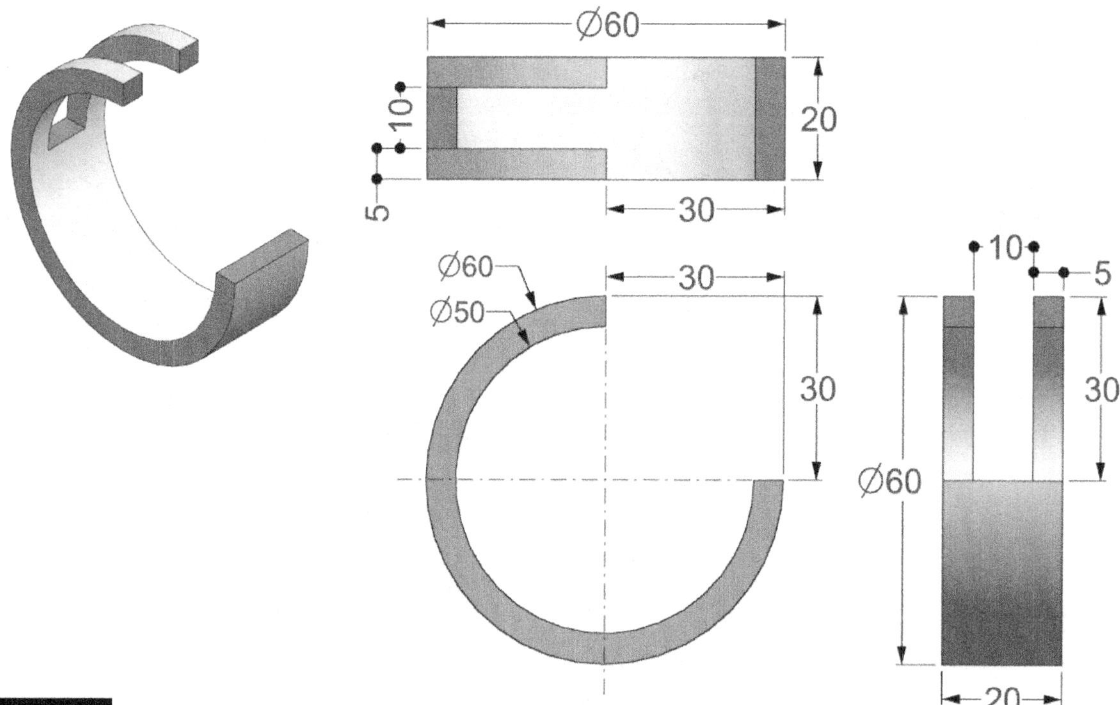
Ø60
10
5
20
30
Ø60
Ø50
30
30
30
Ø60
10
5
20

EX-46
2X Ø16
2X R30
2X Ø12
R15
20
80
30
120
60
35
30
60
45
30
20
50
30

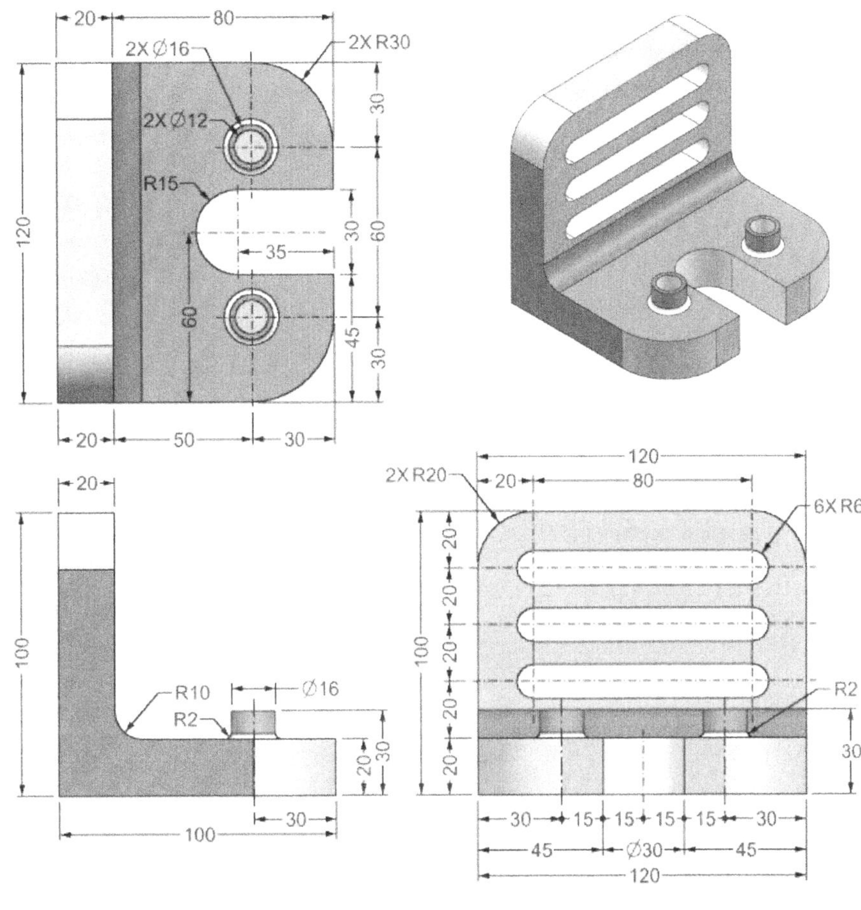

20
100
R10
R2
Ø16
20
30
100
30
EX-47
2X R20
120
20
80
6X R6
20
20
100
20
20
R2
30
30
15
15
15
30
45
Ø30
45
120

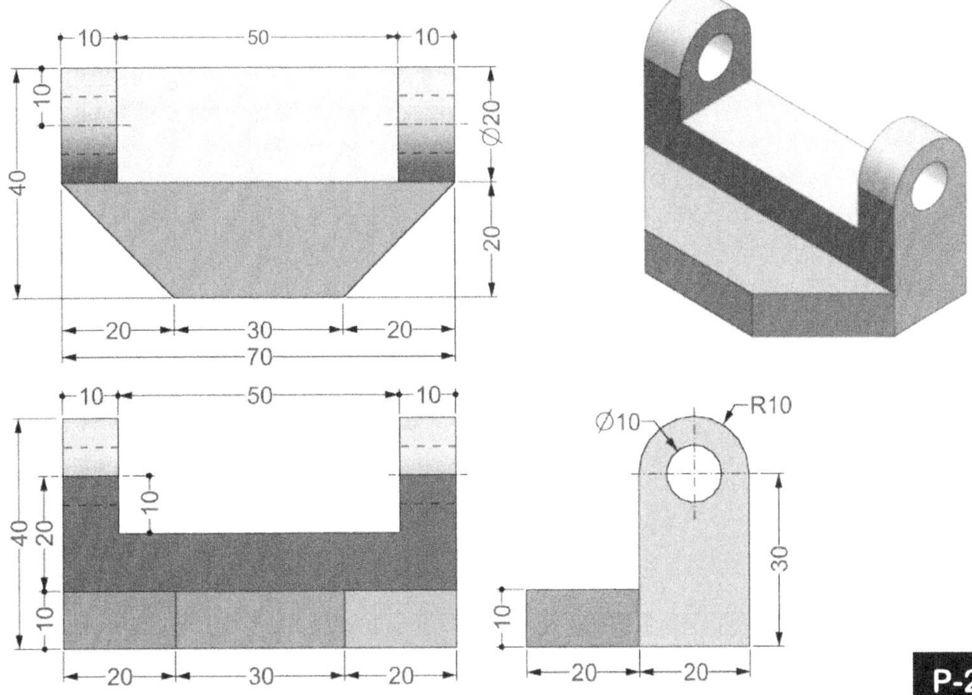

10
50
10
10
Ø20
40
20
20
30
20
70
10
50
10
40
20
10
10
2X Ø16
20
30
20
Ø10
R10
30
20
20
P-25

EX-48

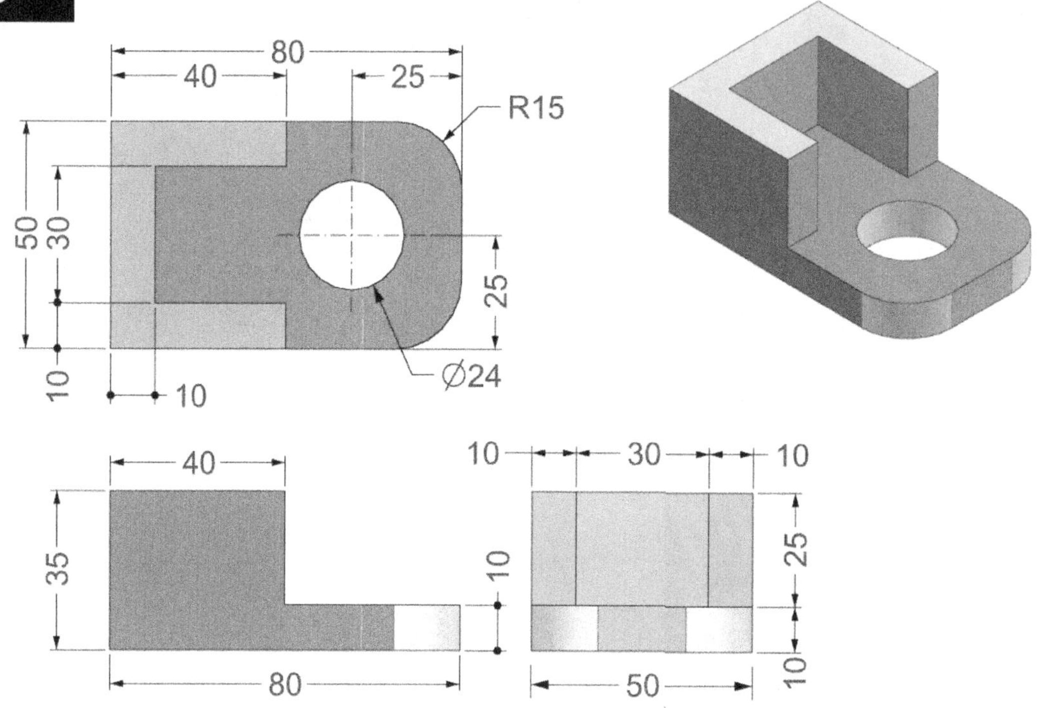
80
40
25
R15
50
30
25
10
10
Ø24
40
35
80
10
10
30
10
10
25
50

EX-49

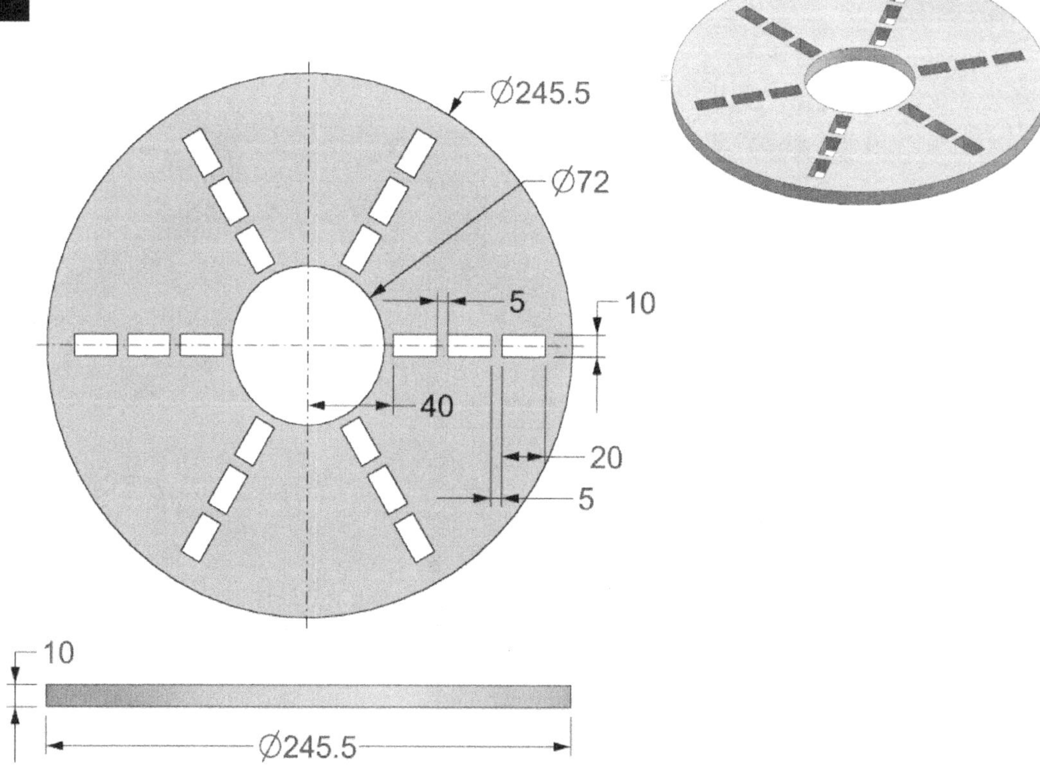
Ø245.5
Ø72
5
10
40
20
5
10
Ø245.5

P-26

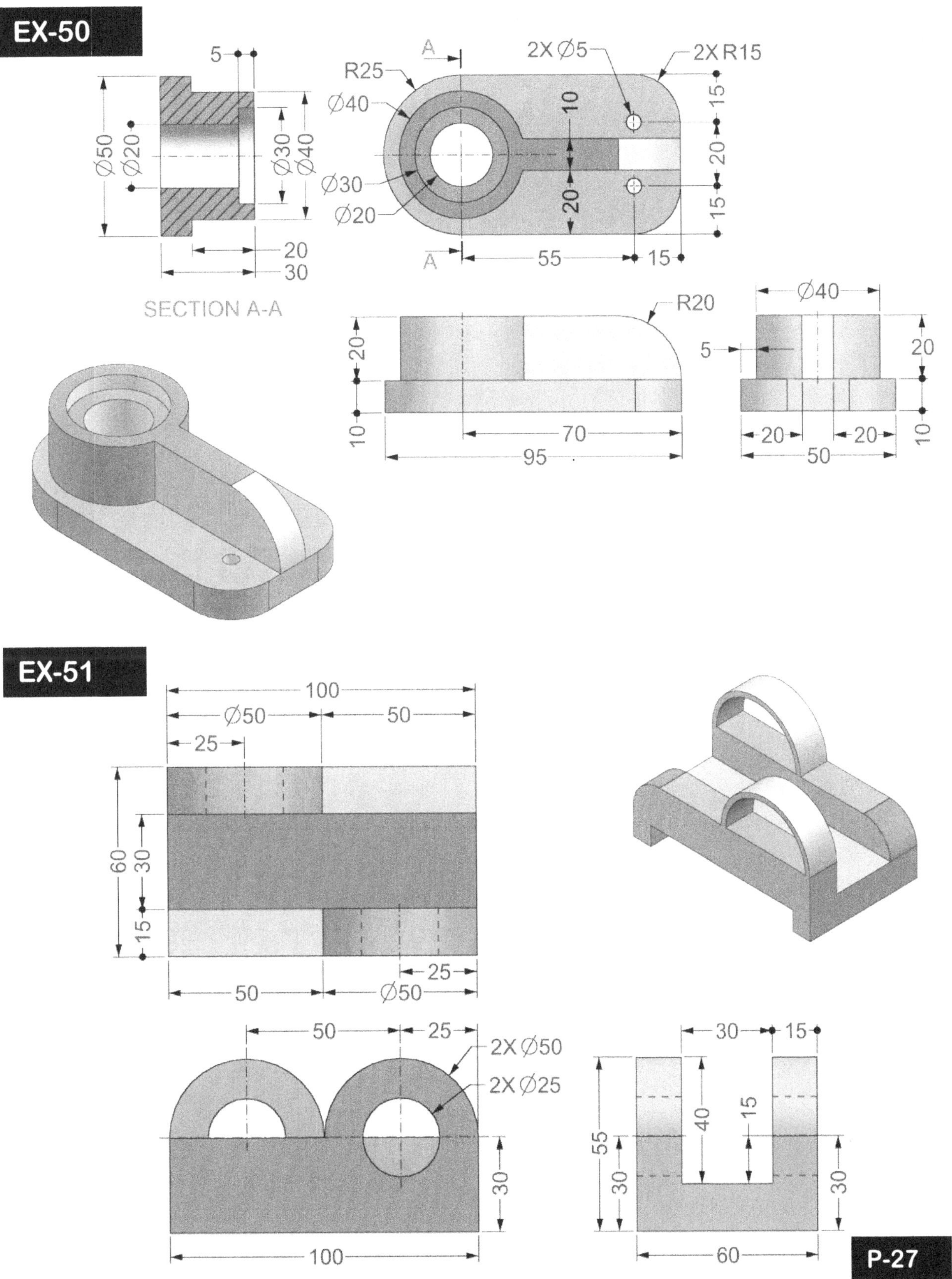

EX-50
5
A
2X Ø5
2X R15
R25
Ø40
10
15
Ø50
Ø20
Ø30
Ø40
15
20
Ø30
Ø20
20
15
20
30
55
15
A
SECTION A-A
20
R20
Ø40
20
10
5
70
20
20
95
50
10

EX-51
100
Ø50
50
25
60
30
15
25
50
Ø50
50
25
2X Ø50
2X Ø25
30
100
30
15
55
40
15
30
30
60

P-27

EX-52 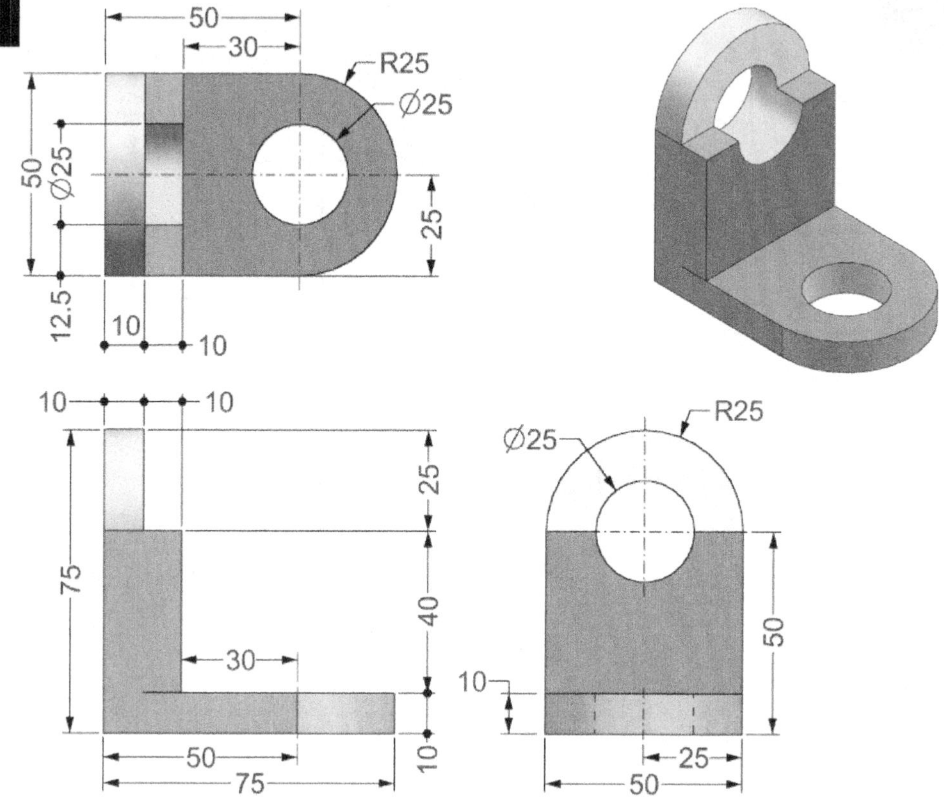
50
30
R25
Ø25
50
Ø25
25
12.5
10
10
10
10
25
75
40
30
50
75
10
Ø25
R25
50
10
25
50
EX-53 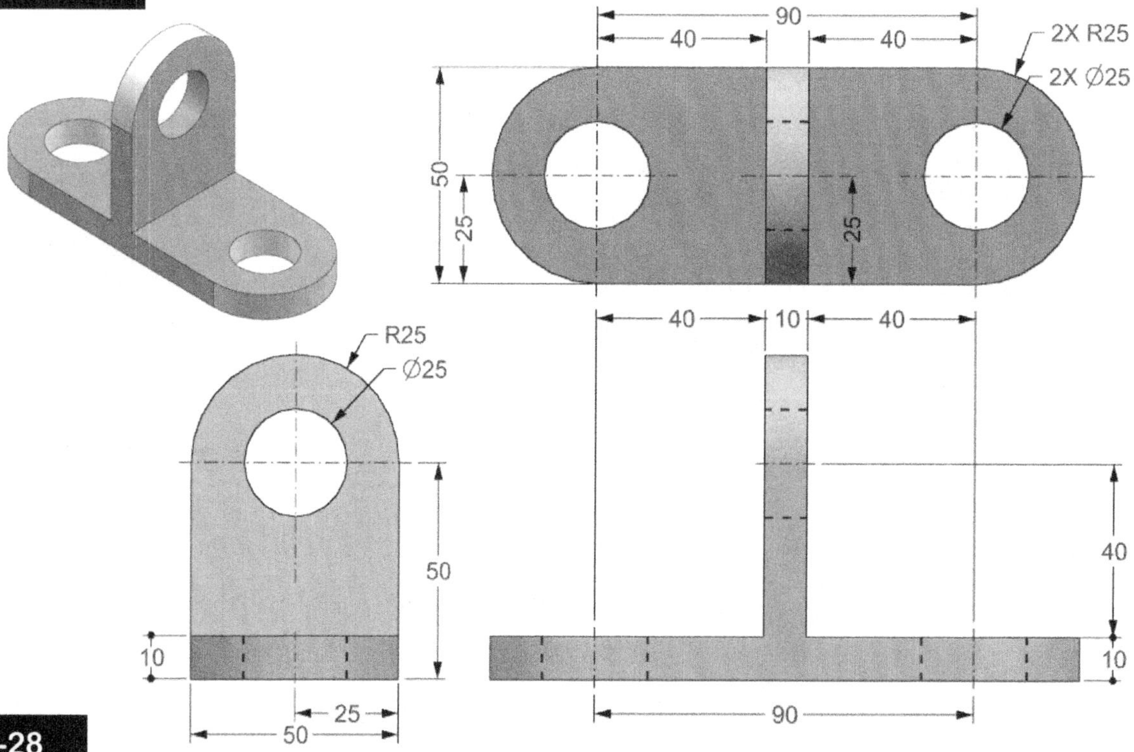
90
40
40
2X R25
2X Ø25
50
25
25
40
10
40
R25
Ø25
50
40
10
25
50
10
90
P-28

EX-54
100 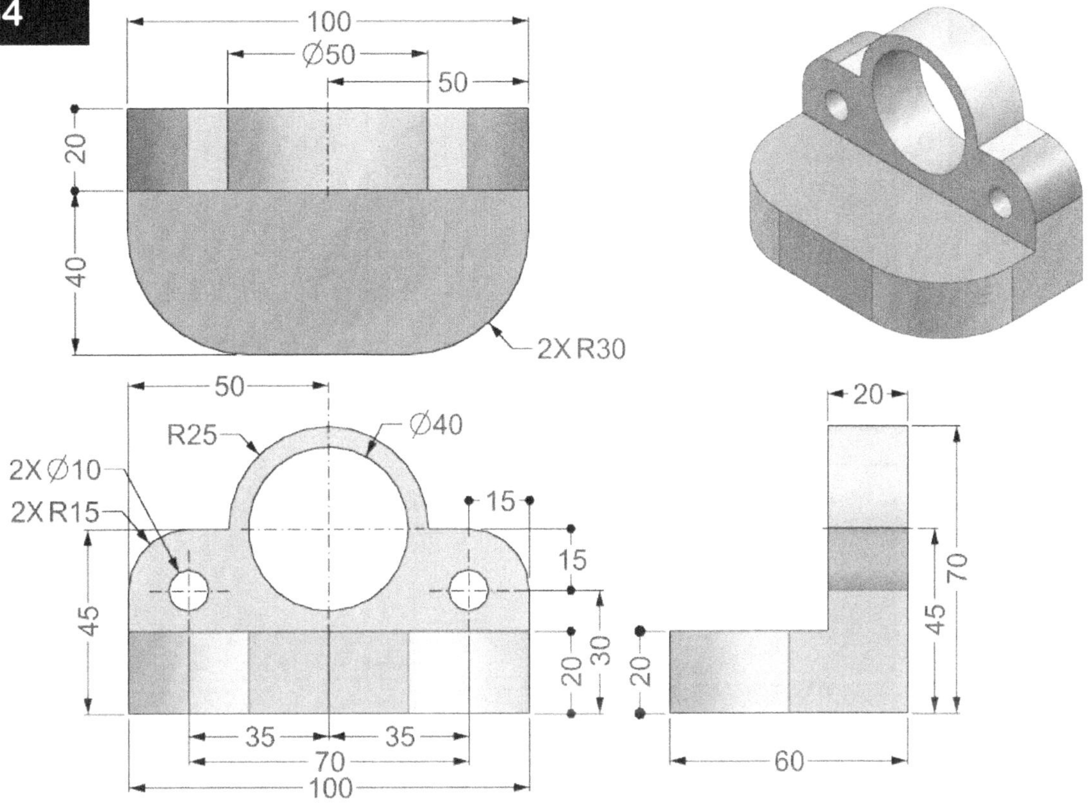
Ø50
50
20
40
2X R30
50
R25
Ø40
2X Ø10
2X R15
15
15
45
30
20
35
35
70
100
20
70
45
20
60
EX-55
100 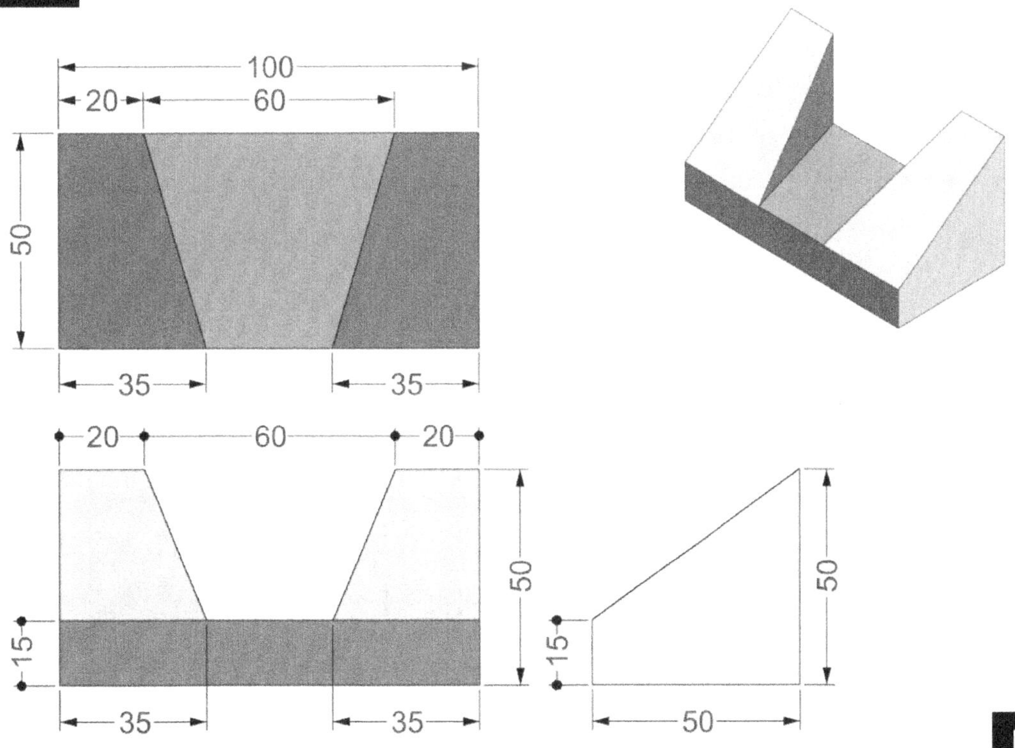
20
60
50
35
35
20
60
20
50
50
15
35
35
50
15
50
50
P-29

EX-56
50
10
30
50
35
20
10
20
40
10 X 45°
10
20
30
10
10
20
40
50
10
25
15 X 45°
30
10
15
10
10
30
50
EX-57
60
50
10
20
40
20
10
10
10
10
50
Ø16
10
10
15
20
10
10
2X R10
10
10
15
40
10
10
10
10
10
10
25
10
60
20
20
40
P-30

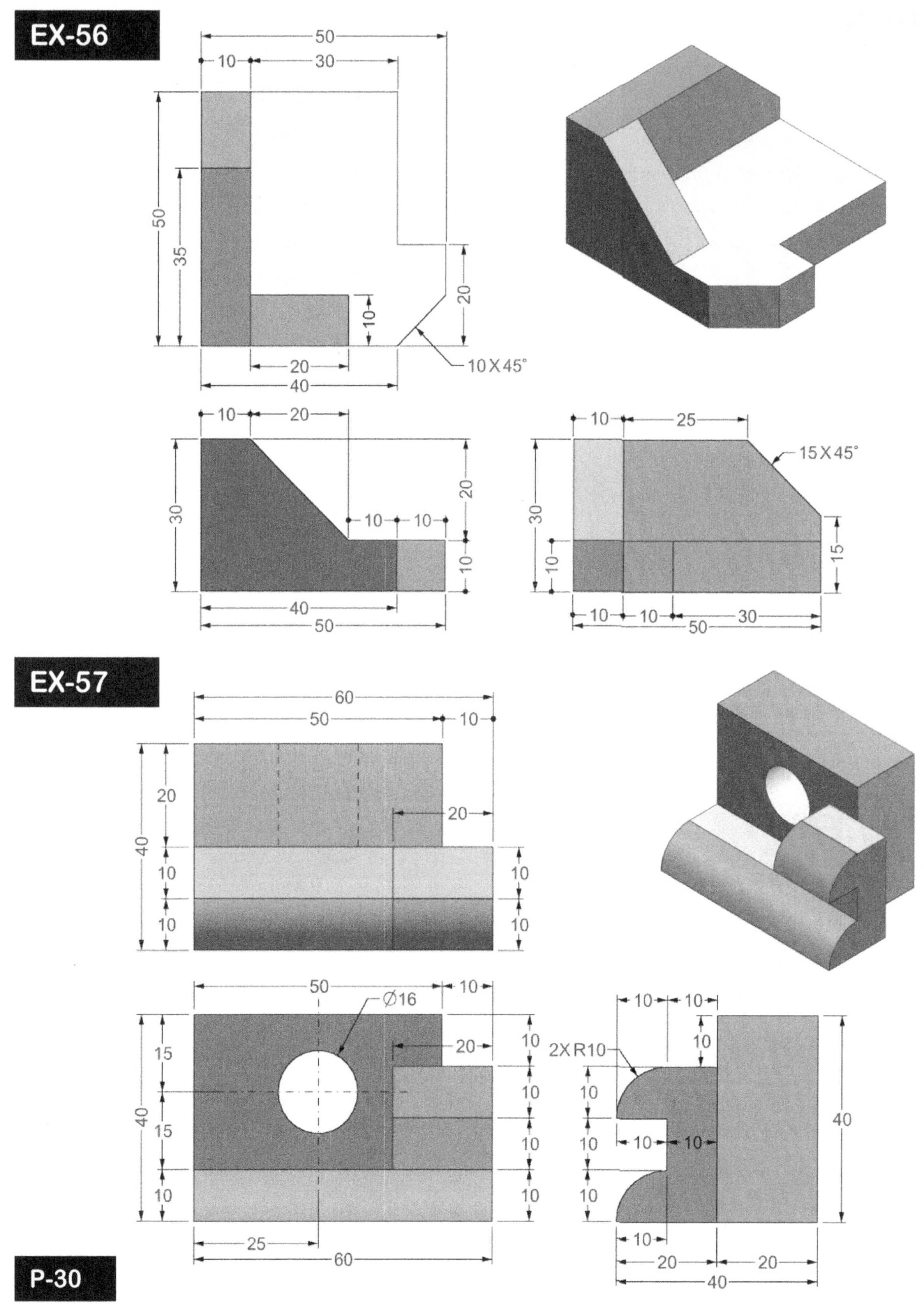

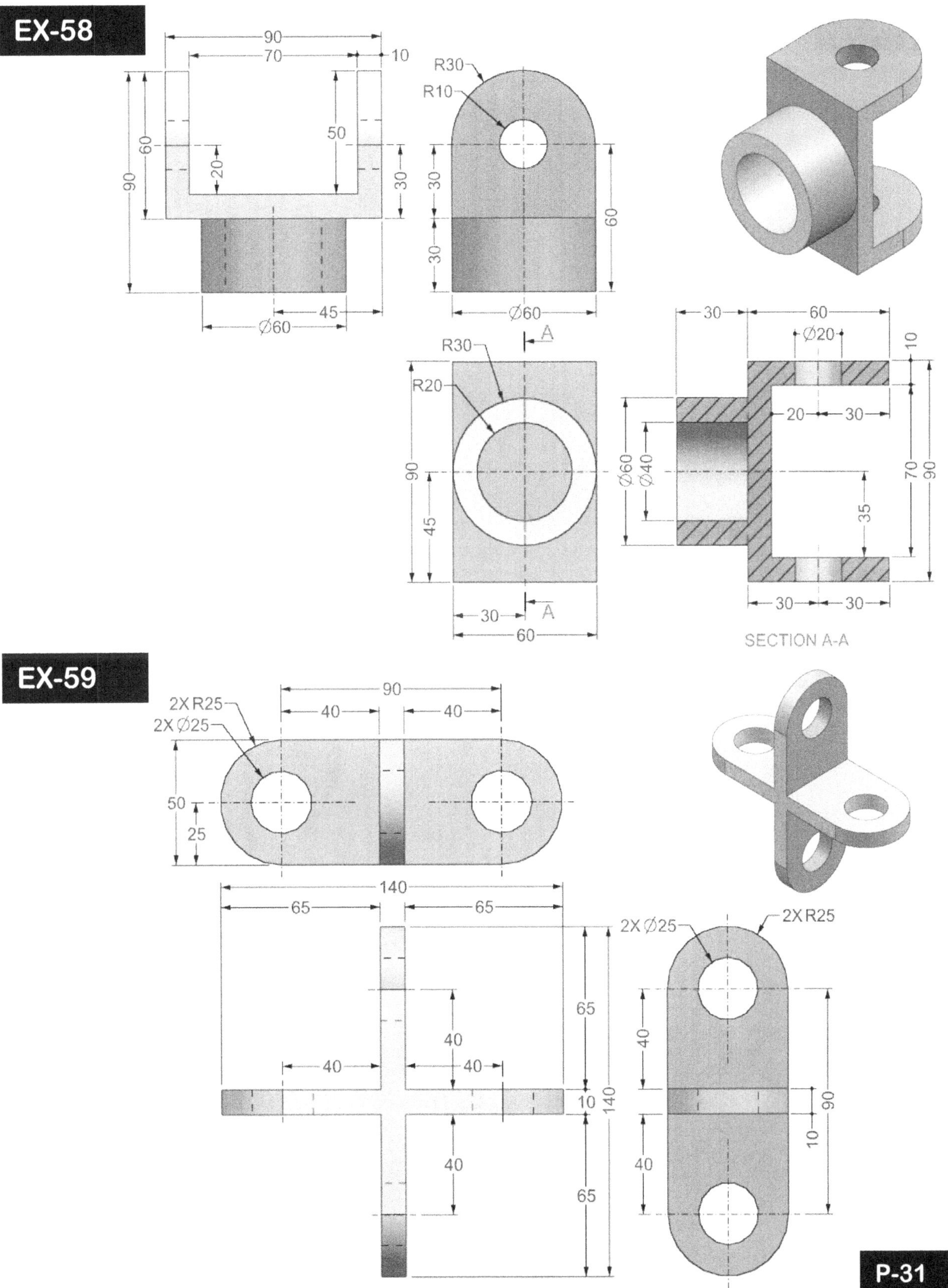

EX-58
90
70
10
R30
R10
60
50
20
30
90
30
30
60
60
45
Ø60
Ø60
A
R30
R20
90
45
30
60
A
30
60
Ø20
10
20
30
70
90
Ø60
Ø40
35
30
30
SECTION A-A
EX-59
2X R25
2X Ø25
90
40
40
50
25
140
65
65
65
40
40
40
40
65
10
140
2X Ø25
2X R25
40
90
10
40
P-31

EX-60
Ø50
22.5
2X Ø10
15
25
60
43.9
10
15
25
50
45
95
Ø50
Ø40
R4
R10
10
R10
40
155
130
10
45
45
R10
R10
30
55
10
22.5
10
22.5
100
60
80
140
100
155
85.4
34.6
10
25
40
60
EX-61
Ø120
20
50
R3
Ø50
R2
10 10 10
Ø50
Ø70
14 14
PCD Ø90
Ø70
Ø50
R60
Ø30
6X Ø10
66
66
132
14
14
A
66
66
66
132
A
132
Ø70
Ø50
Ø30
Ø10
10 10 10
R2
Ø50
Ø30
Ø10
R3
50
100
20
45
45
90
Ø120
SECTION A-A
Ø120
PCD Ø90
6X Ø10
ON PCD 90
Ø30
66
14
14
14
132
66
14 14
66
66
132
P-32

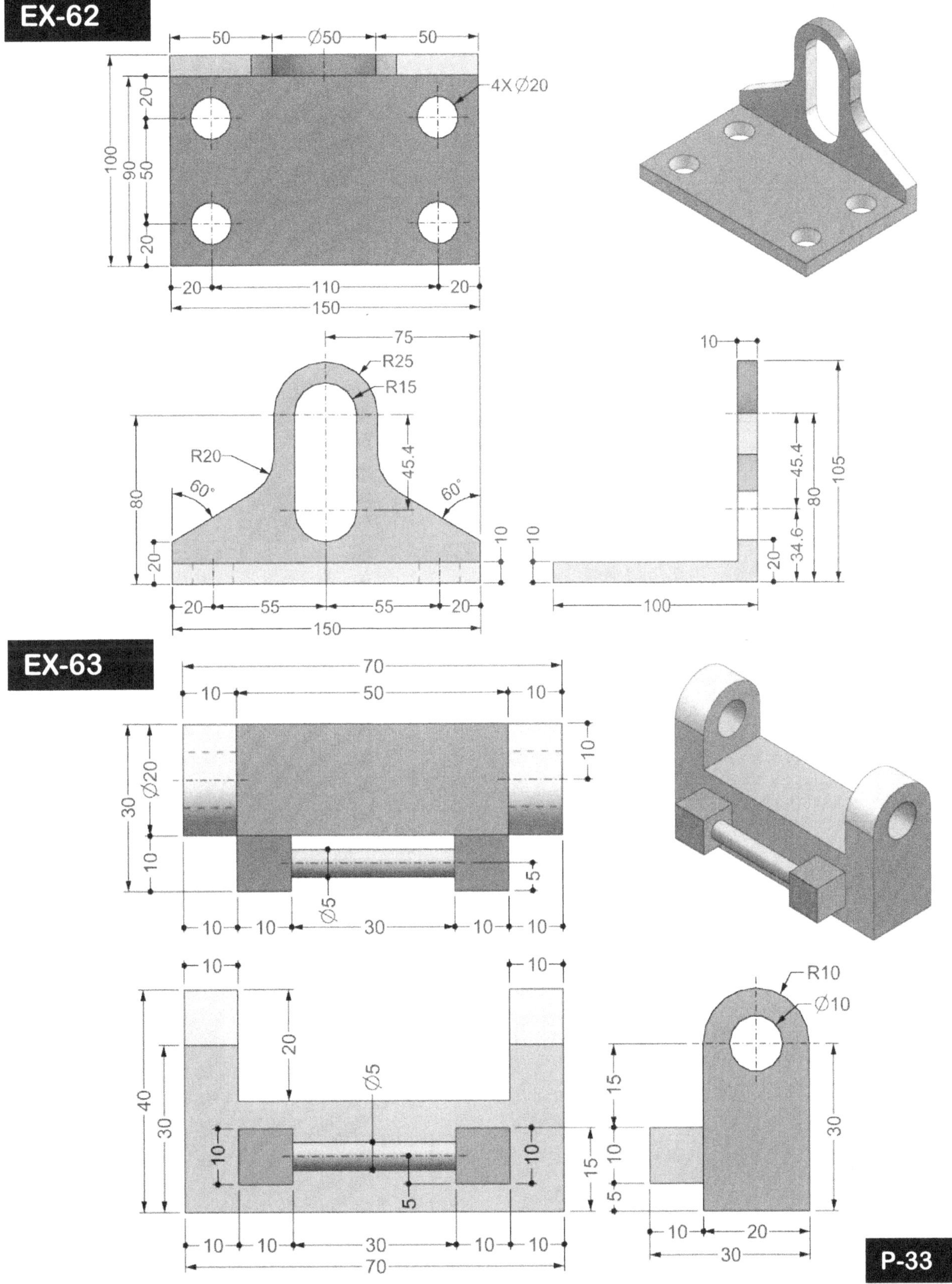

EX-62
50
Ø50
50
4X Ø20
100
90
50
20
20
20
110
20
150
75
R25
R15
R20
45.4
60°
60°
80
20
10
20
55
55
20
150
10
10
45.4
34.6
20
80
105
100
EX-63
70
10
50
10
30
Ø20
10
10
5
10
10
30
10
10
Ø5
10
10
20
40
30
Ø5
10
10
15
15
5
10
10
30
10
10
70
R10
Ø10
15
30
10
5
10
20
30
P-33

EX-64

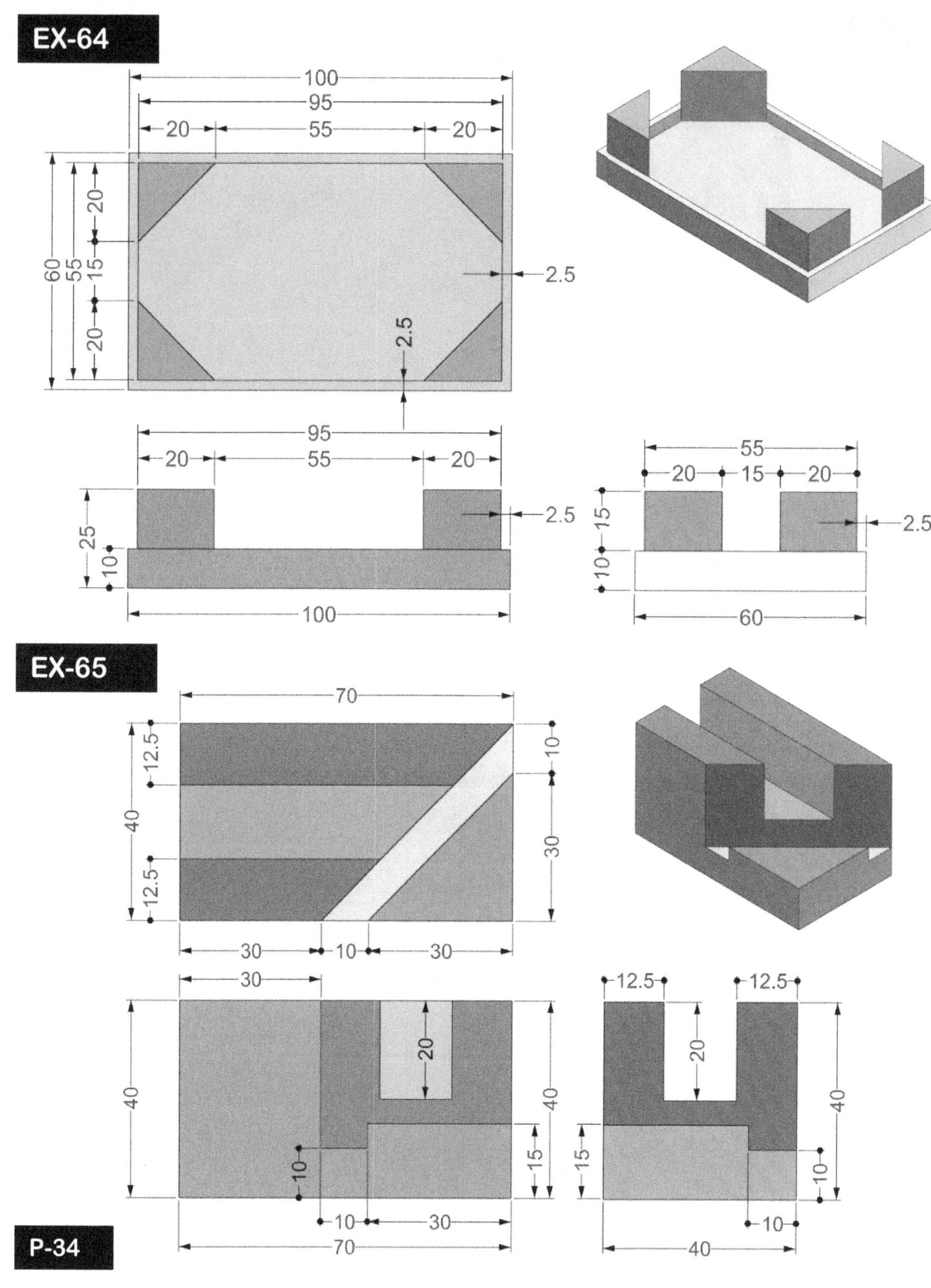

EX-65
P-34

EX-66
30 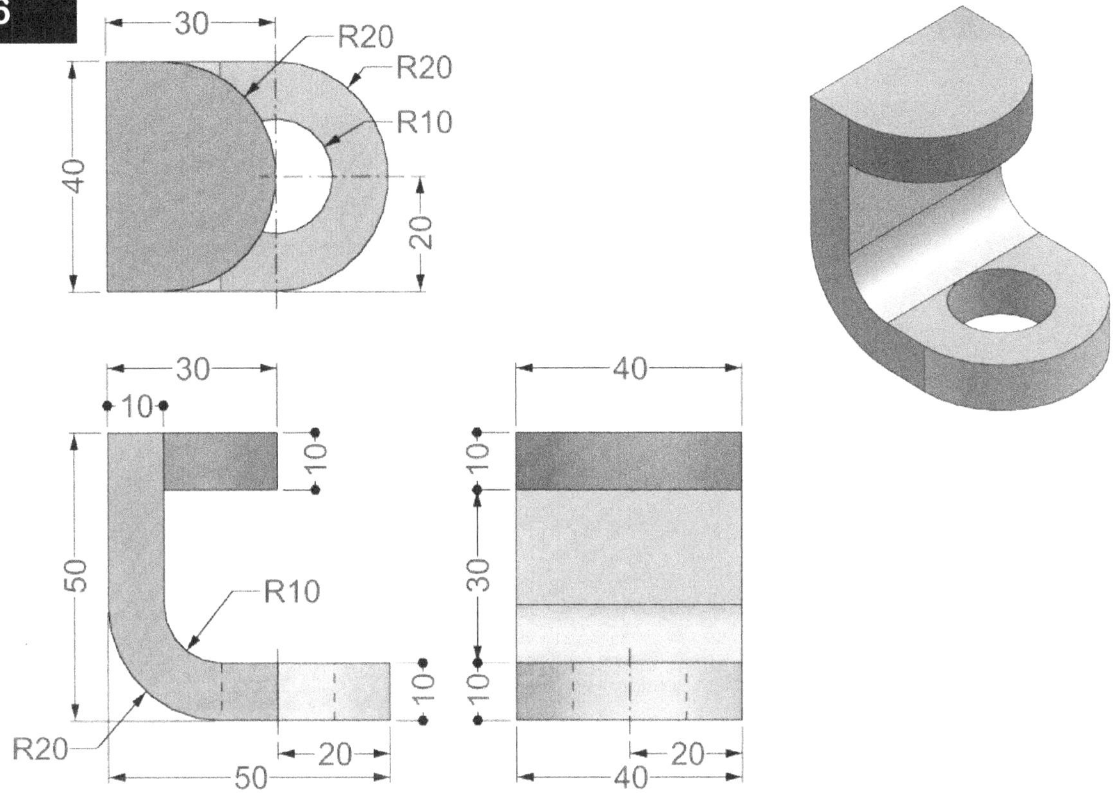
R20
R20
R10
40
20
30
10
10
50
R10
R20
20
50
40
10
30
10
20
40

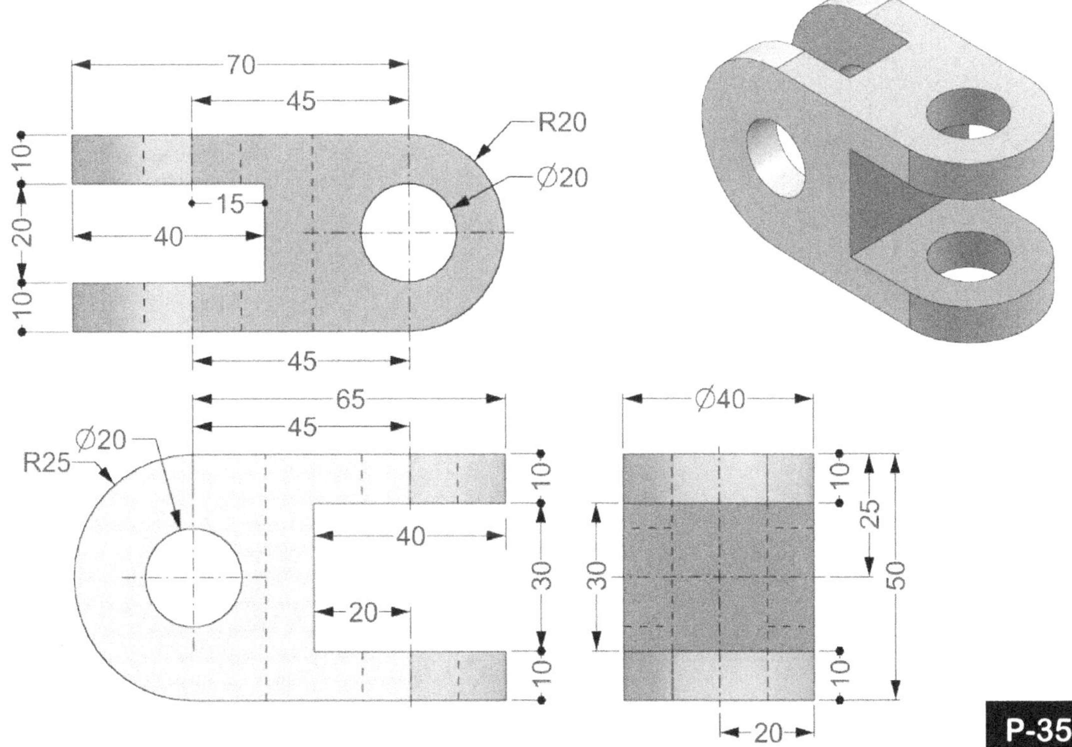

EX-67
70
45
R20
Ø20
10
10
20
15
40
10
45
65
45
Ø20
R25
40
30
20
30
10
Ø40
10
25
30
50
10
20
P-35

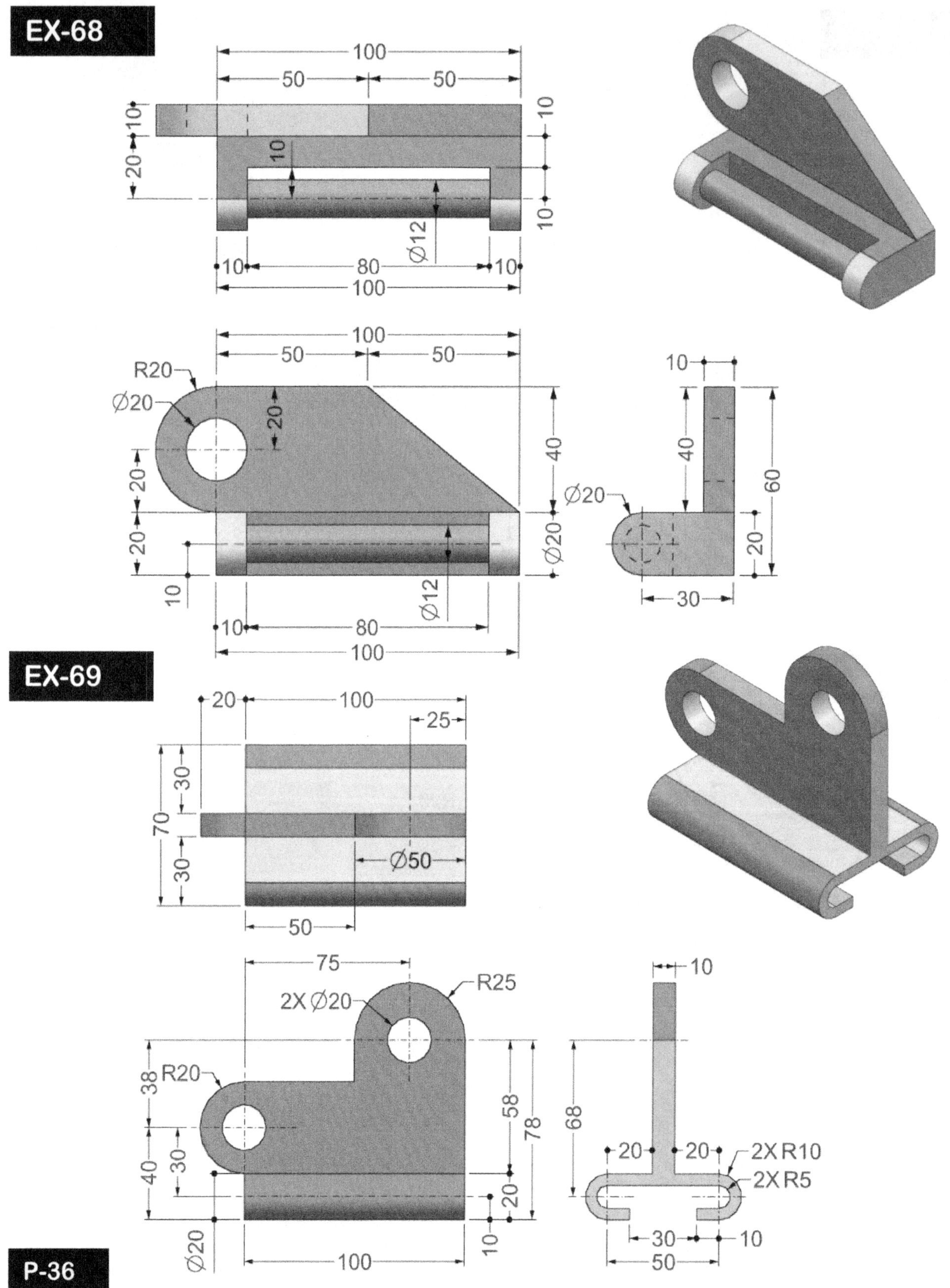

EX-68
EX-69
P-36
100
50
50
20
10
10
10
10
10
80
10
100
Ø12
R20
Ø20
20
40
20
20
20
10
10
80
10
100
Ø20
Ø20
Ø12
10
40
60
20
30
Ø20
20
100
25
30
70
30
Ø50
50
75
2X Ø20
R25
R20
38
58
78
40
30
20
10
Ø20
100
10
68
20
20
2X R10
2X R5
30
10
50

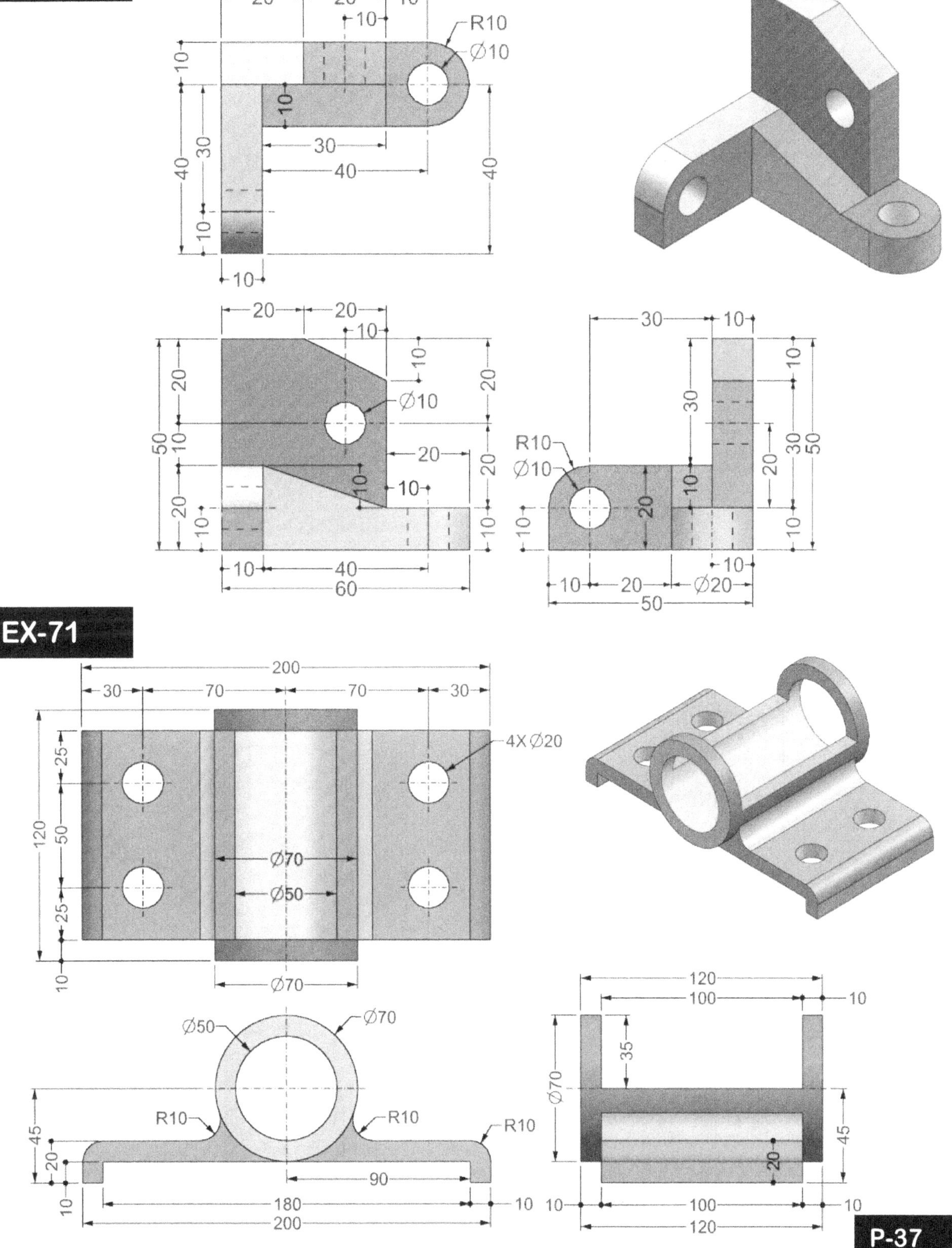

EX-70
EX-71
P-37

EX-72

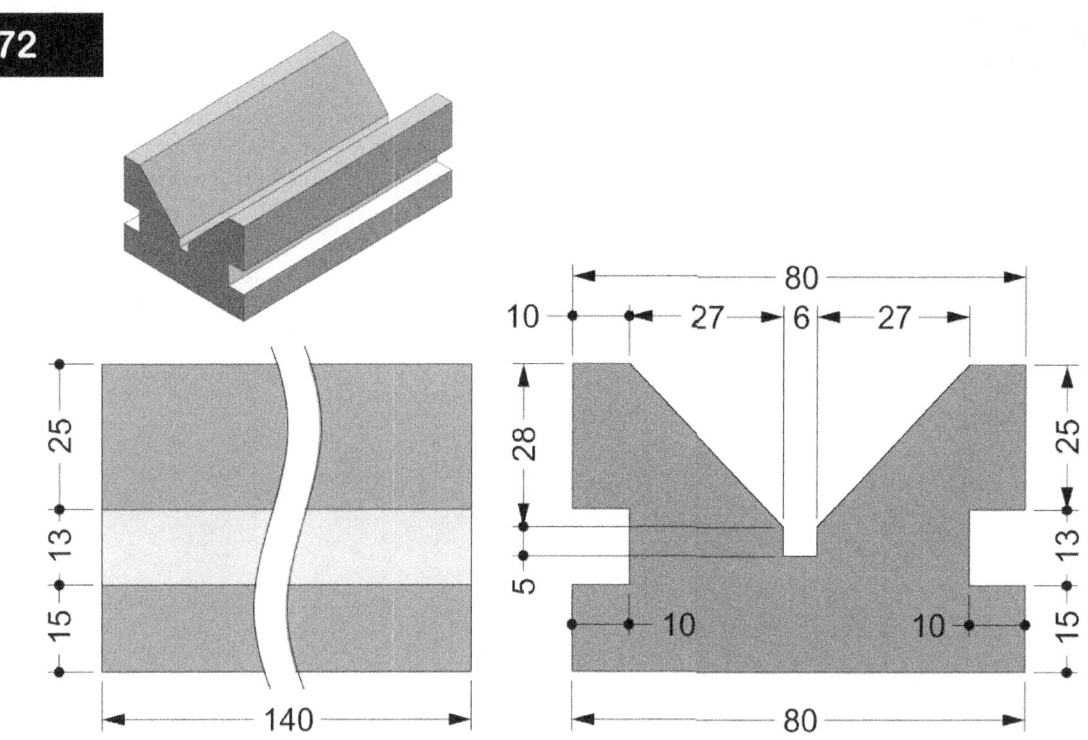

25
13
15
140
80
10
27
6
27
28
5
10
10
25
13
15
80

EX-73 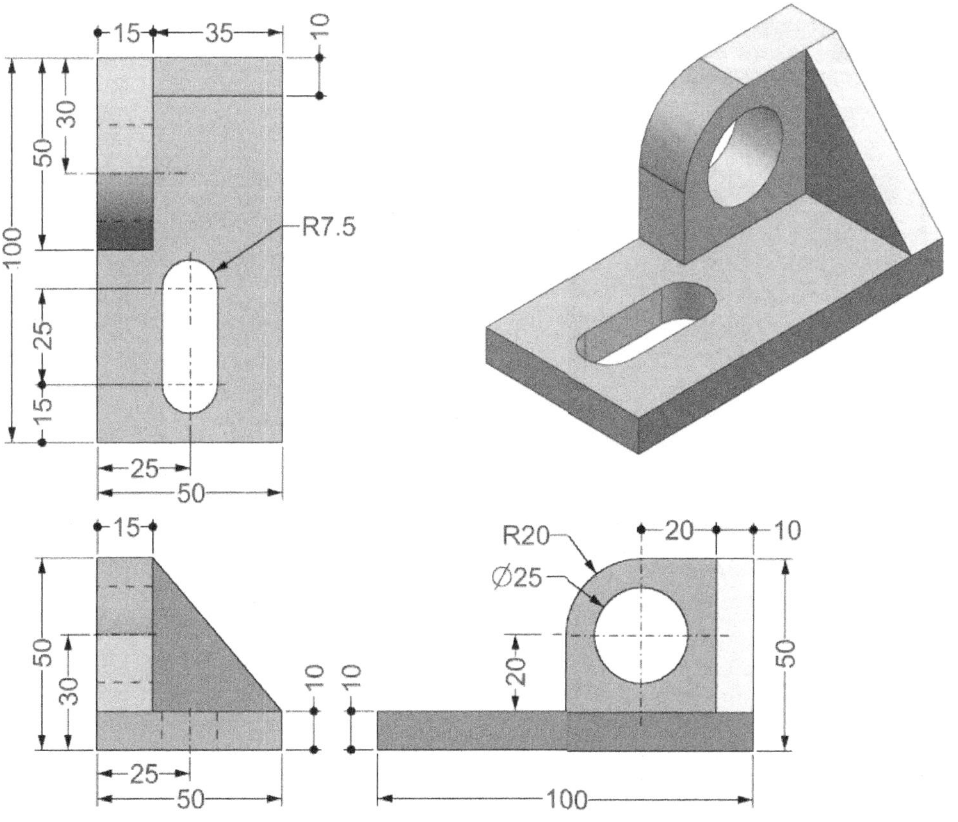
15
35
10
30
50
100
R7.5
25
15
25
50
15
50
30
10
10
25
50
R20
Ø25
20
10
20
50
100

P-38

EX-74 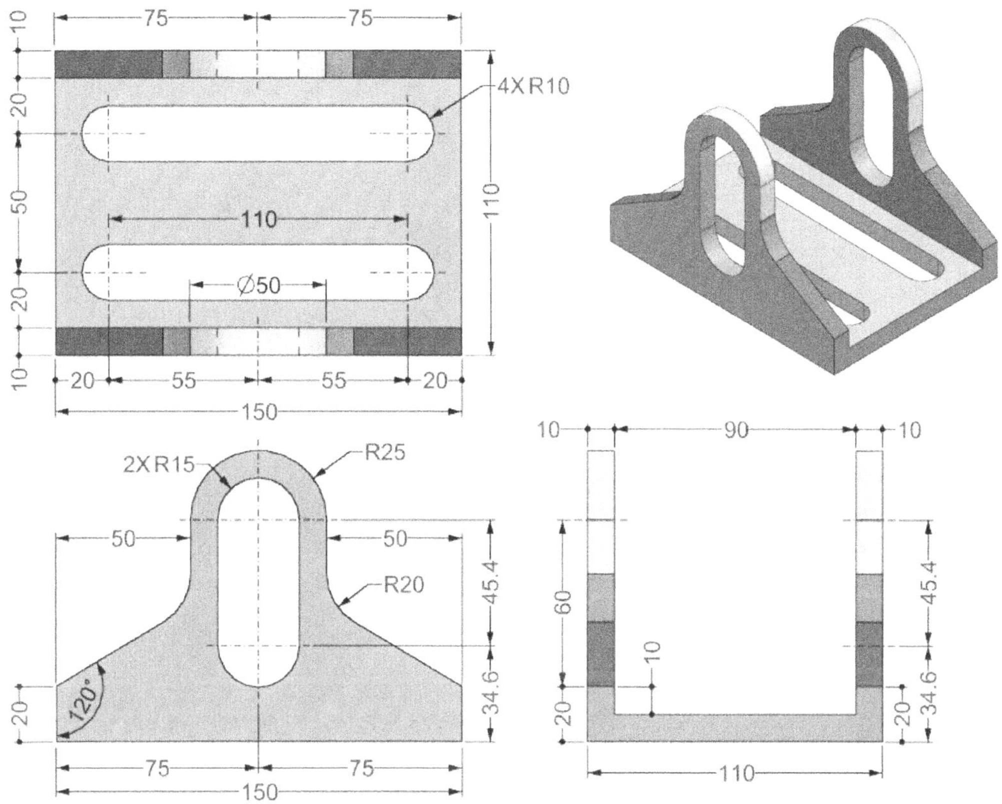
4XR10
75
75
10
20
50
20
10
110
110
Ø50
20
55
55
20
150
2XR15
R25
R20
50
50
45.4
34.6
120°
20
75
75
150
10
90
10
60
10
45.4
20
34.6
20
110
EX-75
40 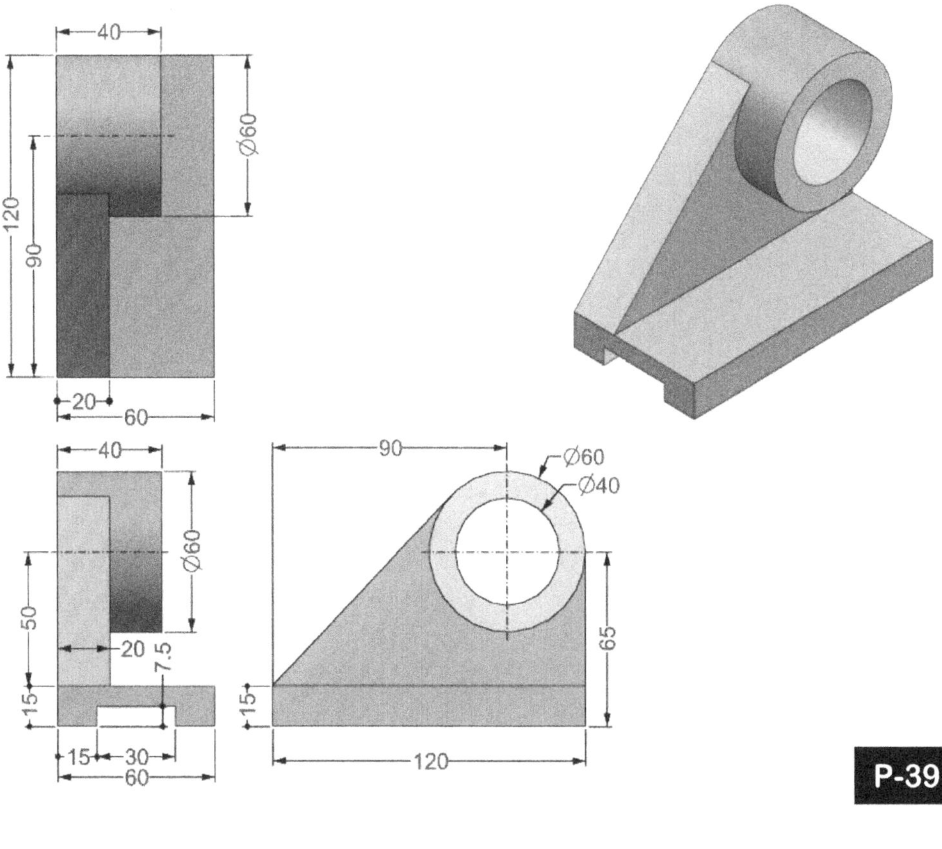
Ø60
120
90
20
60
40
Ø60
50
20
7.5
15
15
30
60
90
Ø60
Ø40
65
15
120
P-39

EX-76
120 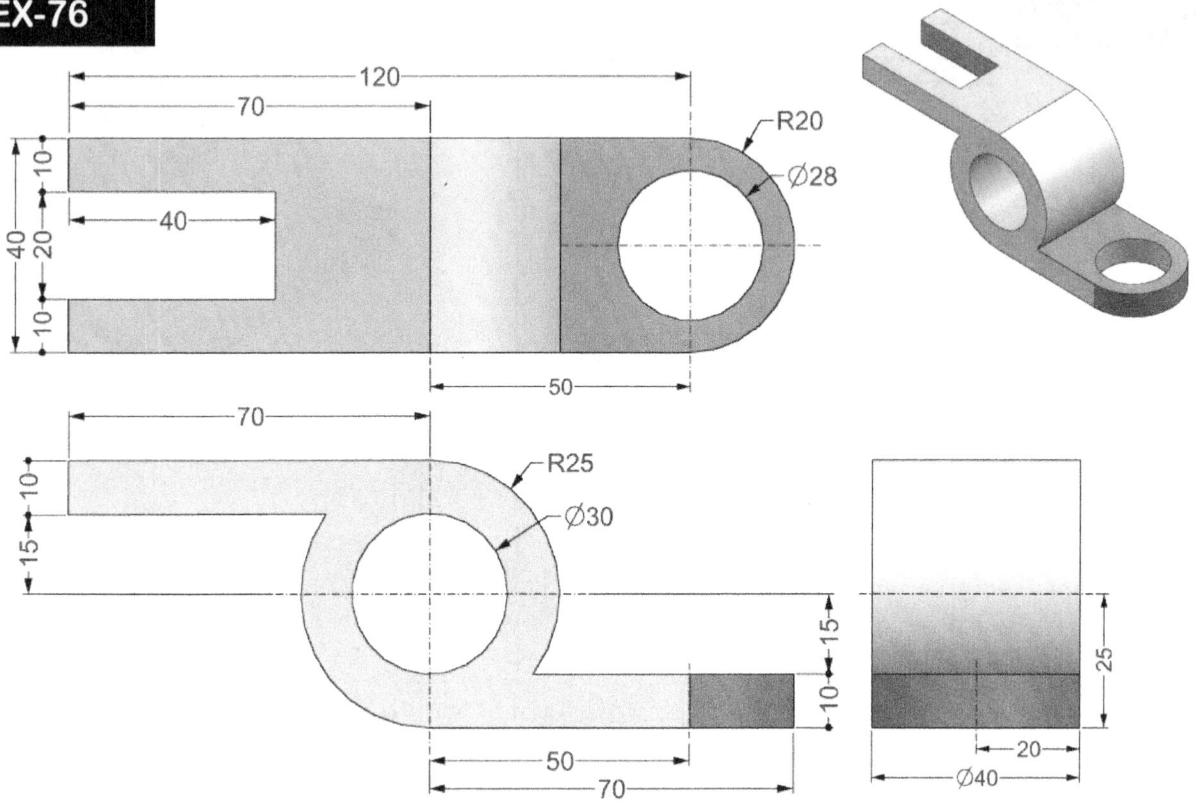
70
R20
Ø28
10
40
20
40
10
50
70
R25
Ø30
10
15
15
50
10
70
25
20
Ø40
EX-77
R15
R25
50
R10
R5
A
A
50
Ø30
R1
30
30
30
20
R4
R2
Ø10
5
SECTION A-A
P-40

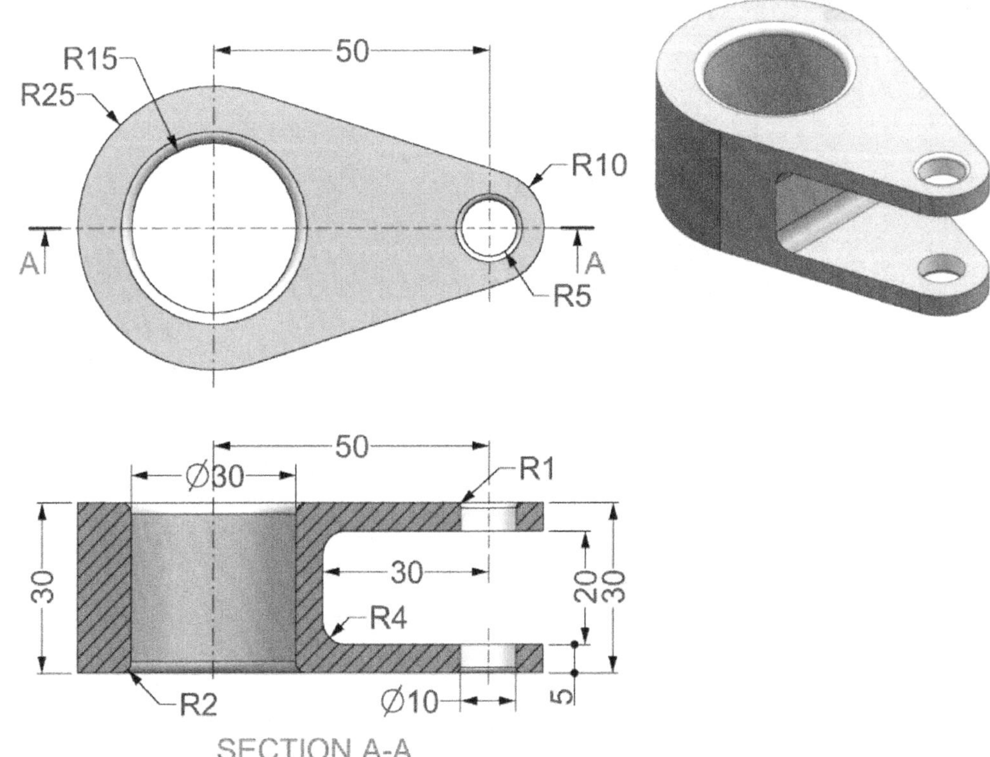

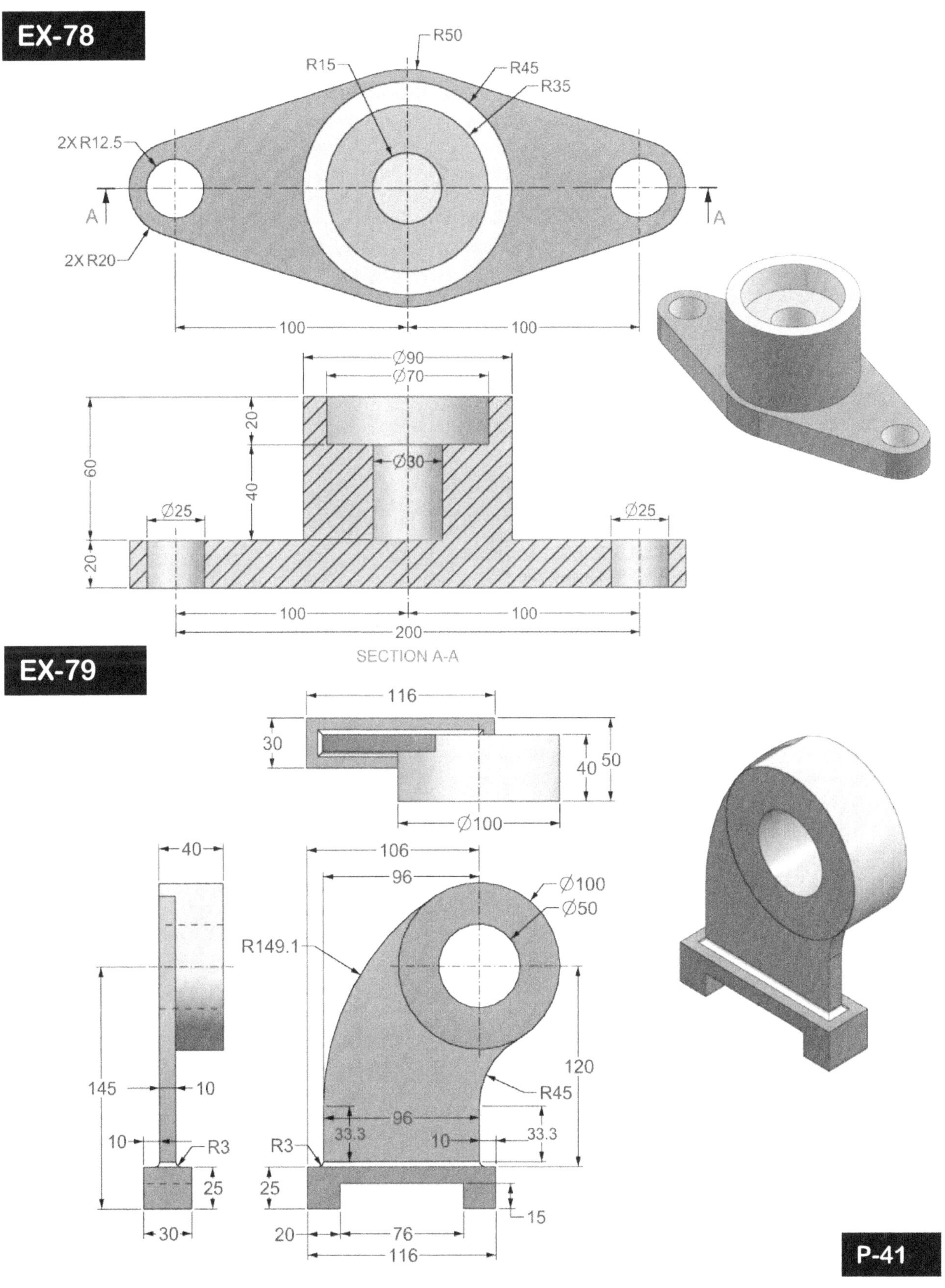

EX-80
6 HOLES,⌀10
ON DIA 32 PCD
4 HOLES,⌀8.6
ON DIA 54 PCD
⌀70
⌀16
A
A
⌀54
⌀32
CHAMFER 0.5 X 45°
4X ⌀8.6
⌀16
6X ⌀10
10
5
5
SECTION A-A
(SCALE 1:1)

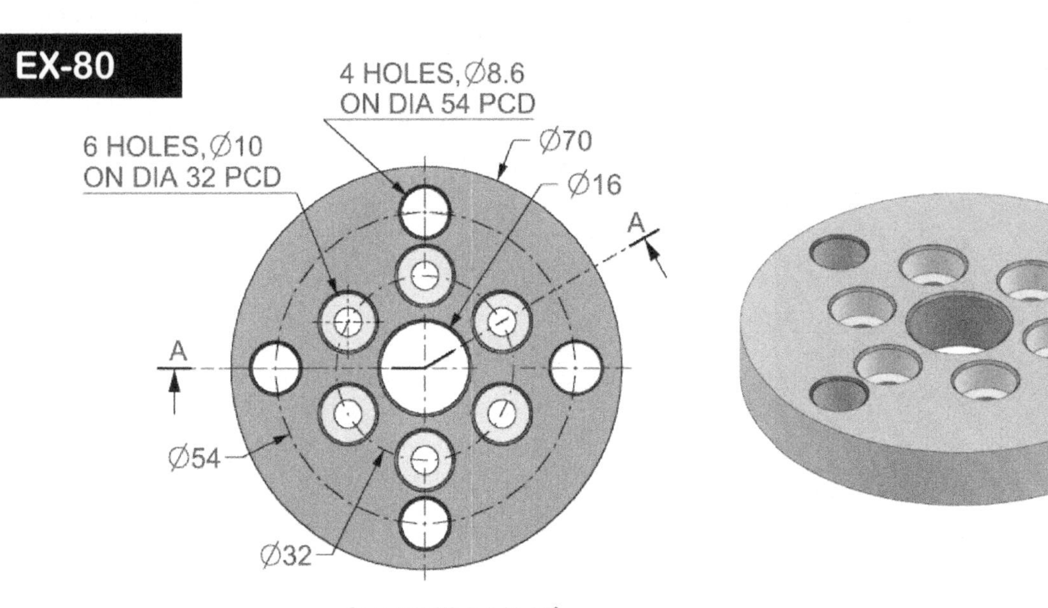

EX-81
10
6X ⌀8.4
207.2
171.6
17.8
87.2
4X R19.4
19.2
106
254
9.6
233.6
254
109.8
190.4
56.4
36.6
38 28
10
60
10
2X R11.6
103.6
147.2

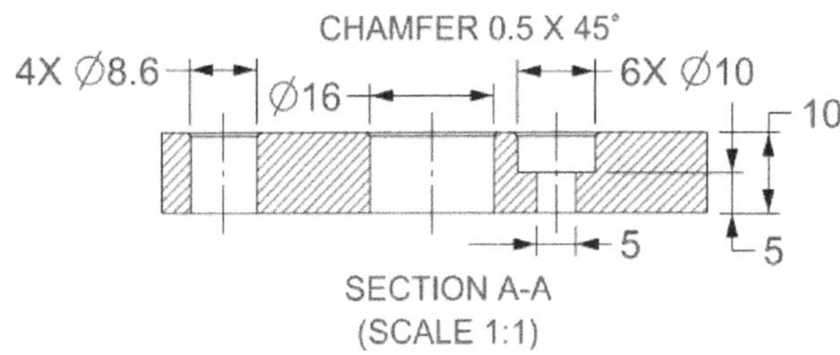

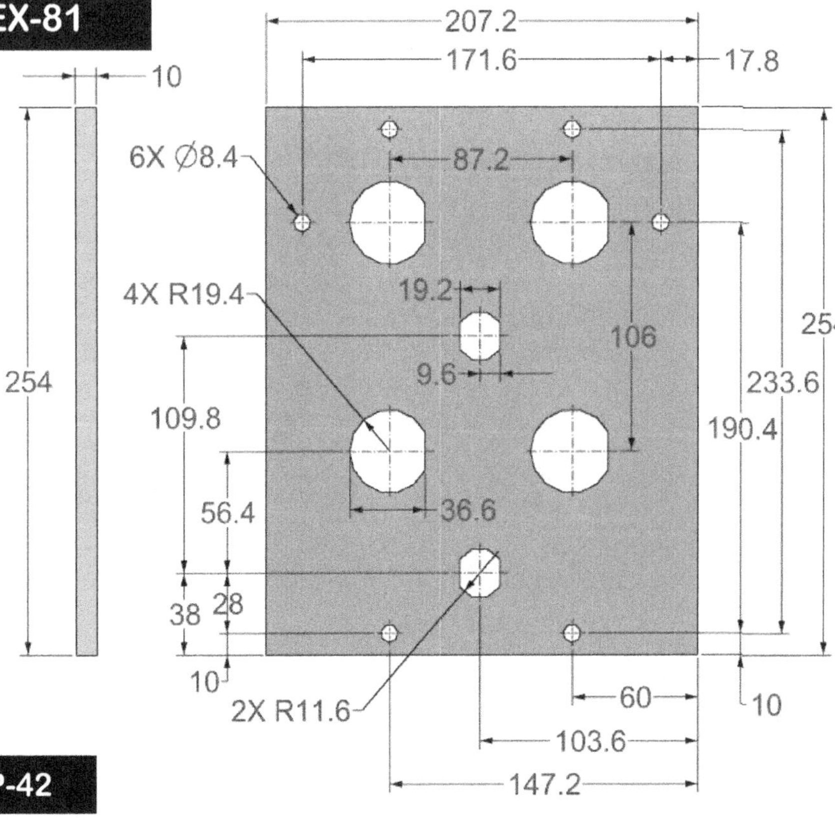

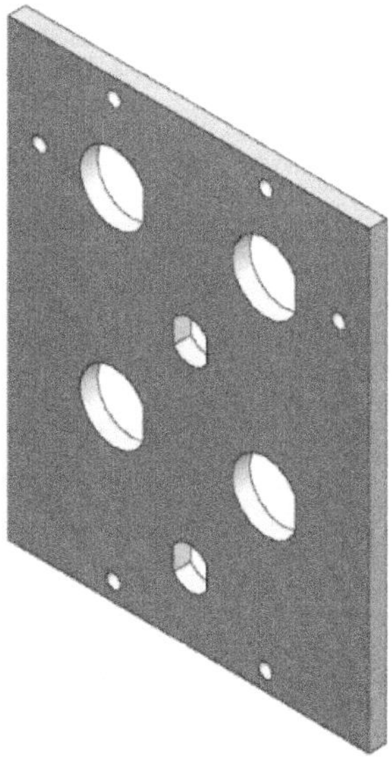

EX-82 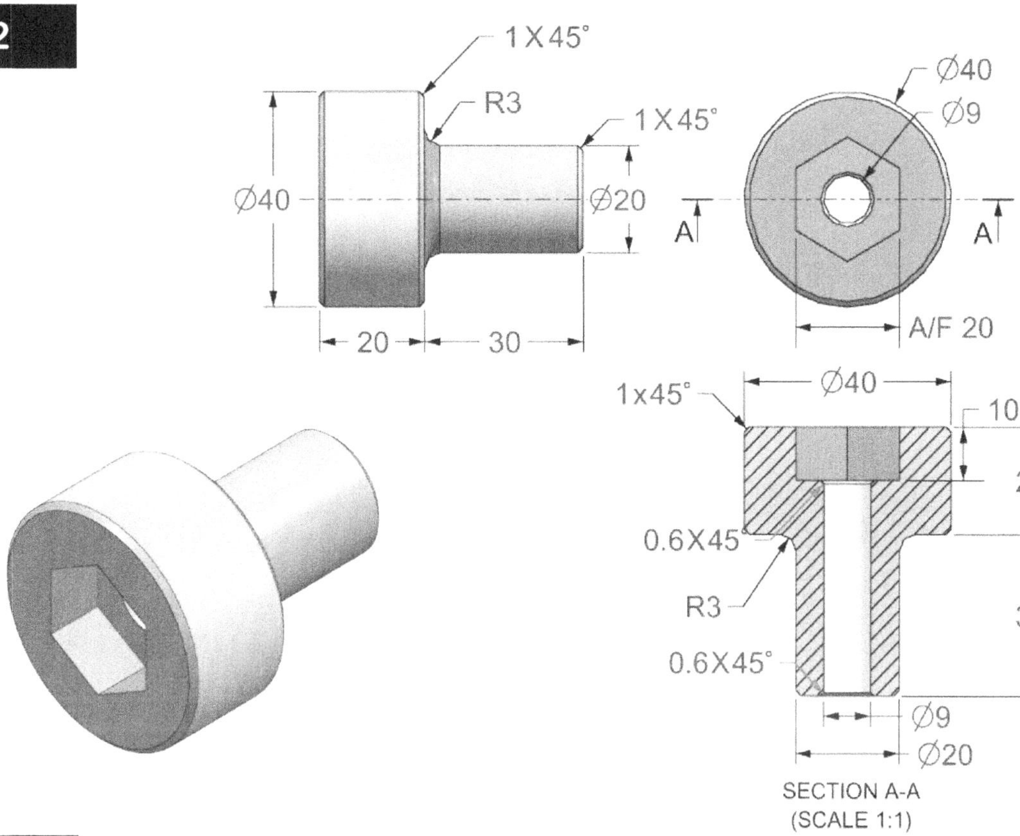
1 X 45°
R3
1 X 45°
Ø40
Ø20
20
30
Ø40
Ø9
A
A
A/F 20
1x45°
Ø40
10
20
0.6X45
R3
30
0.6X45°
Ø9
Ø20
SECTION A-A
(SCALE 1:1)
EX-83 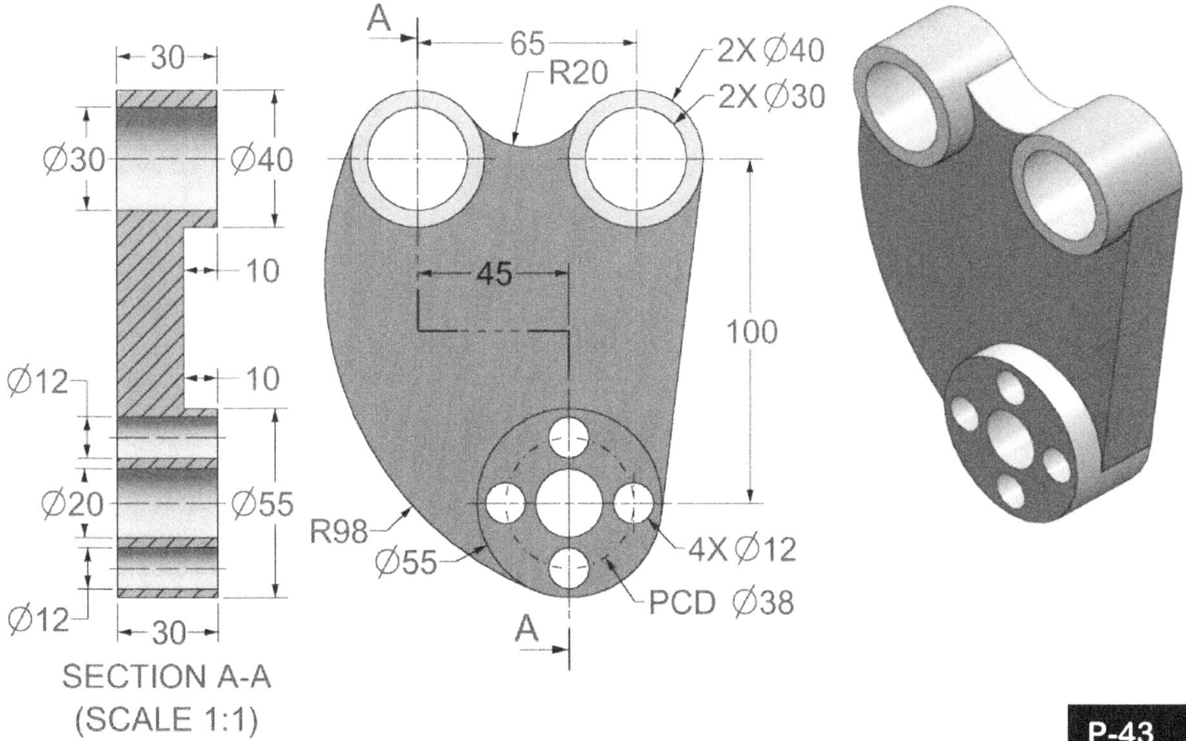
30
A
65
R20
2X Ø40
2X Ø30
Ø30
Ø40
10
45
10
100
Ø12
Ø20
Ø55
R98
4X Ø12
Ø55
PCD Ø38
Ø12
30
A
SECTION A-A
(SCALE 1:1)
P-43

EX-84 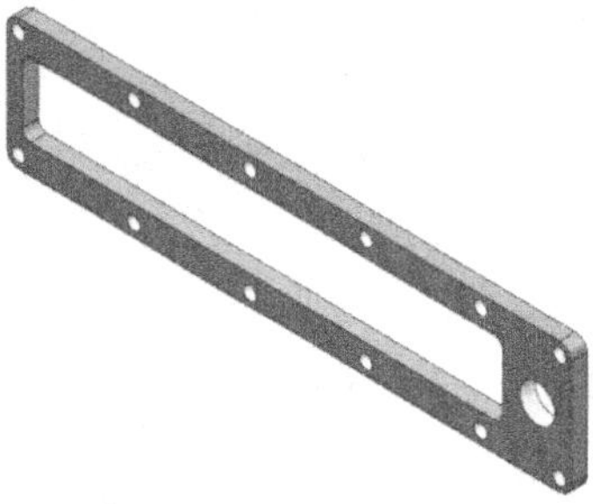 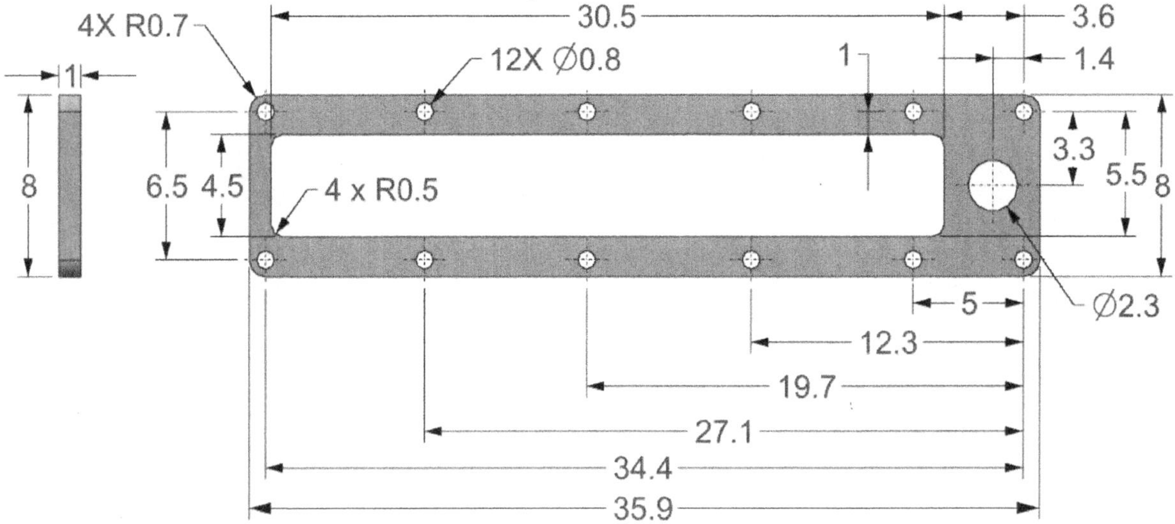
4X R0.7
1
12X ⌀0.8
1
3.6
1.4
30.5
3.3
5.5
8
8
6.5 4.5
4 x R0.5
⌀2.3
5
12.3
19.7
27.1
34.4
35.9
EX-85
30
30
4
4
4
8.5
8.5
40
40
4
20
A
A
8.5
8.5
40
4
4
8
60
4
SECTION A-A
(SCALE 1:1)
P-44

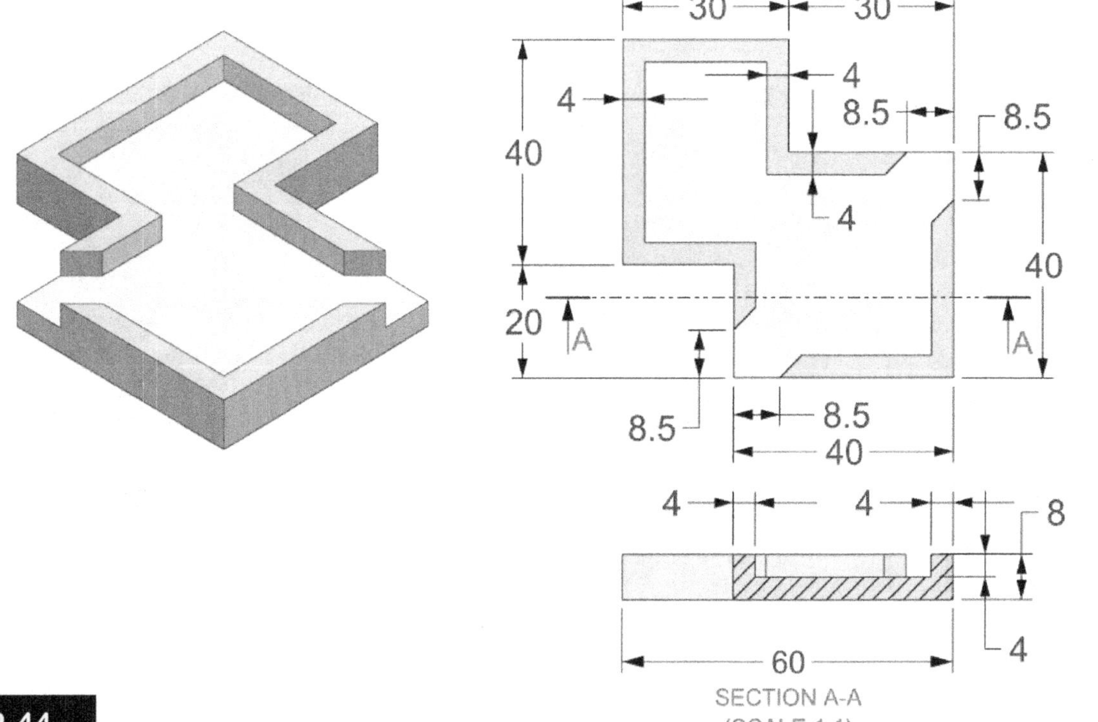

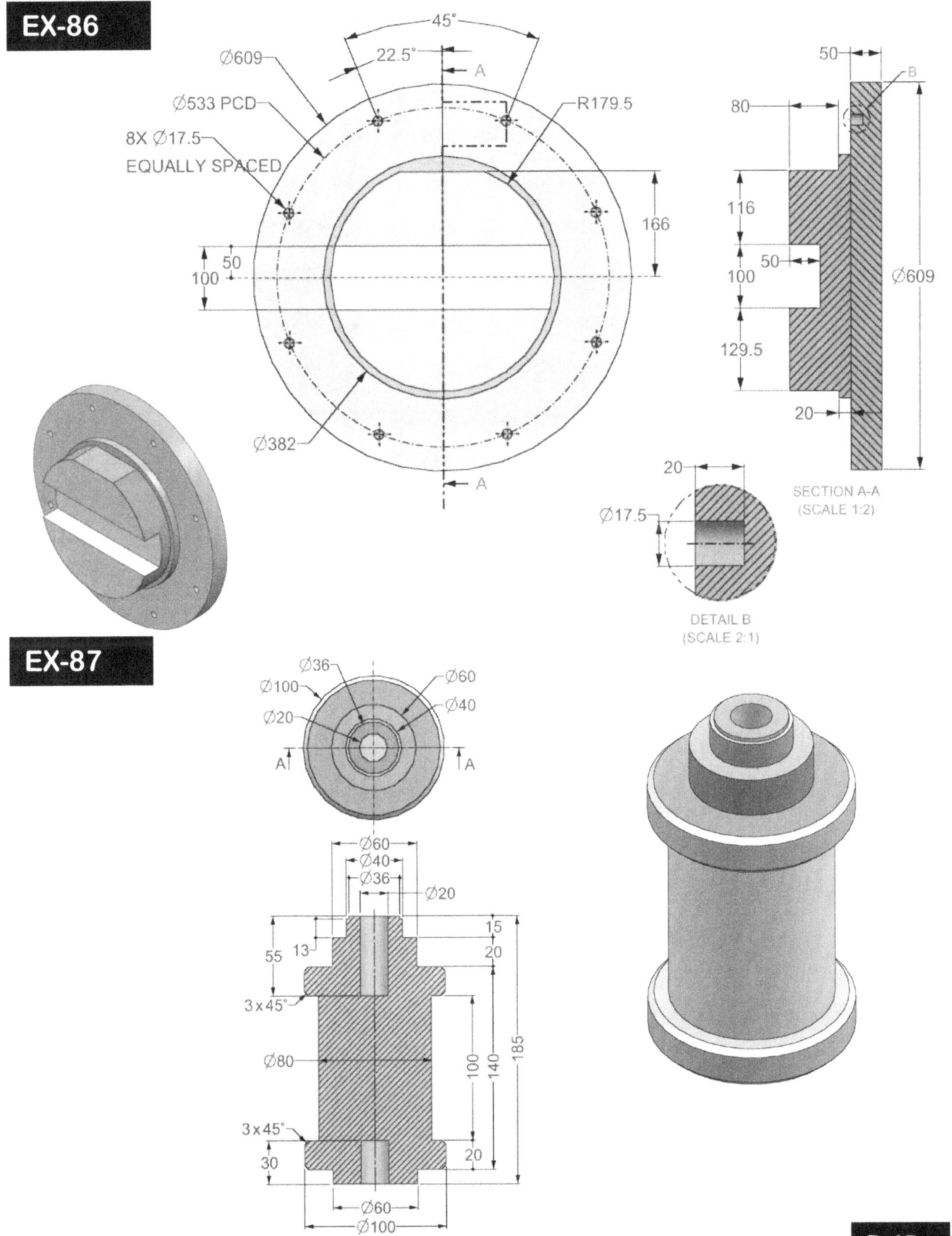

EX-86
EX-87
P-45
Ø609
Ø533 PCD
8X Ø17.5
EQUALLY SPACED
45°
22.5°
A
R179.5
166
50
100
Ø382
A
50
80
116
50
100
129.5
20
B
Ø609
SECTION A-A
(SCALE 1:2)
20
Ø17.5
DETAIL B
(SCALE 2:1)
Ø36
Ø100
Ø20
Ø60
Ø40
A
A
Ø60
Ø40
Ø36
Ø20
15
20
55
13
100
140
185
3 x 45°
Ø80
3 x 45°
30
20
Ø60
Ø100
SECTION A-A

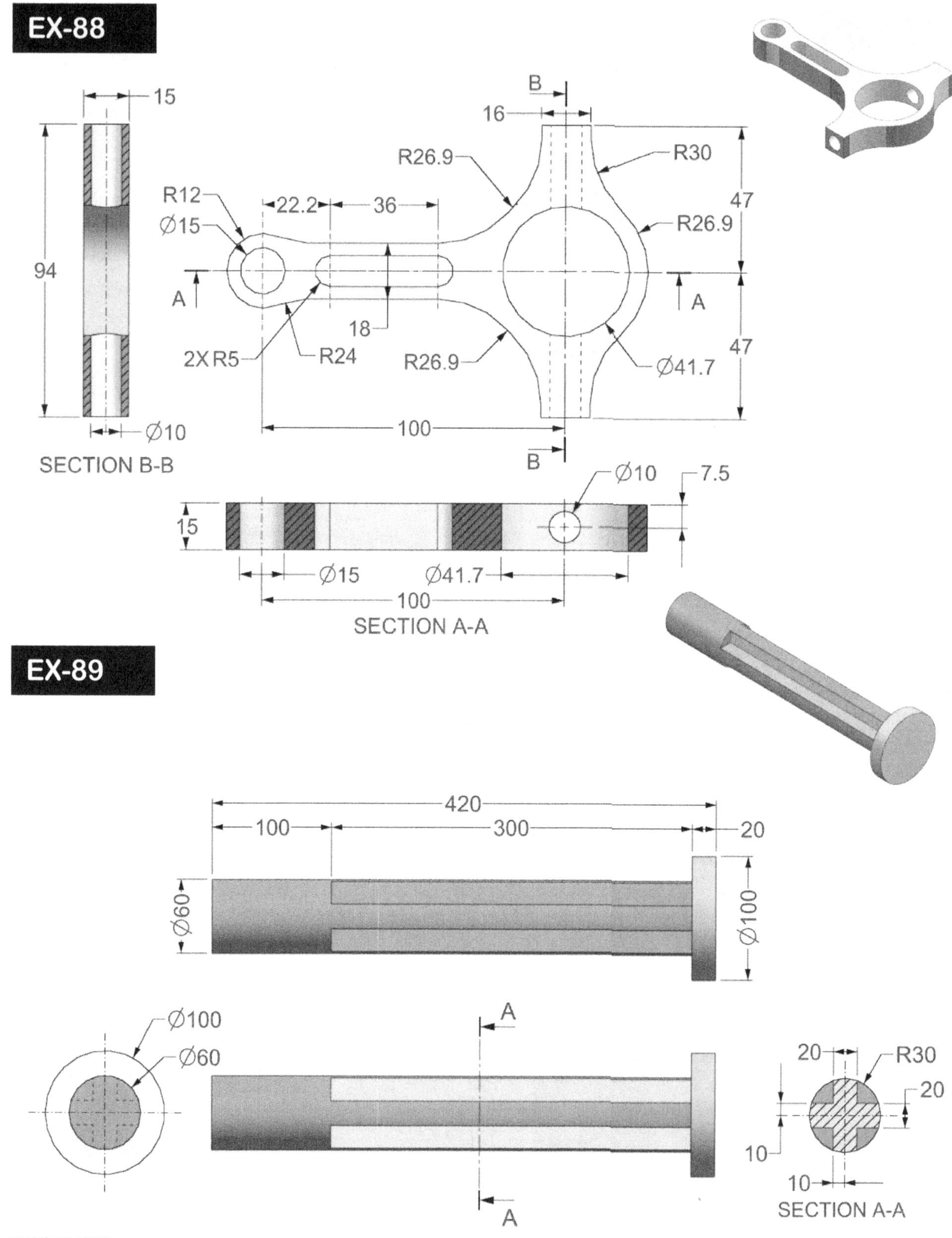

EX-88
15
94
Ø10
SECTION B-B
R12
Ø15
22.2
36
R26.9
16
B
R30
47
R26.9
18
2X R5
R24
R26.9
100
B
Ø41.7
47
A
A
15
Ø10
7.5
Ø15
Ø41.7
100
SECTION A-A
EX-89
420
100
300
20
Ø60
Ø100
Ø100
Ø60
A
A
20
R30
20
10
10
SECTION A-A
P-46

EX-90

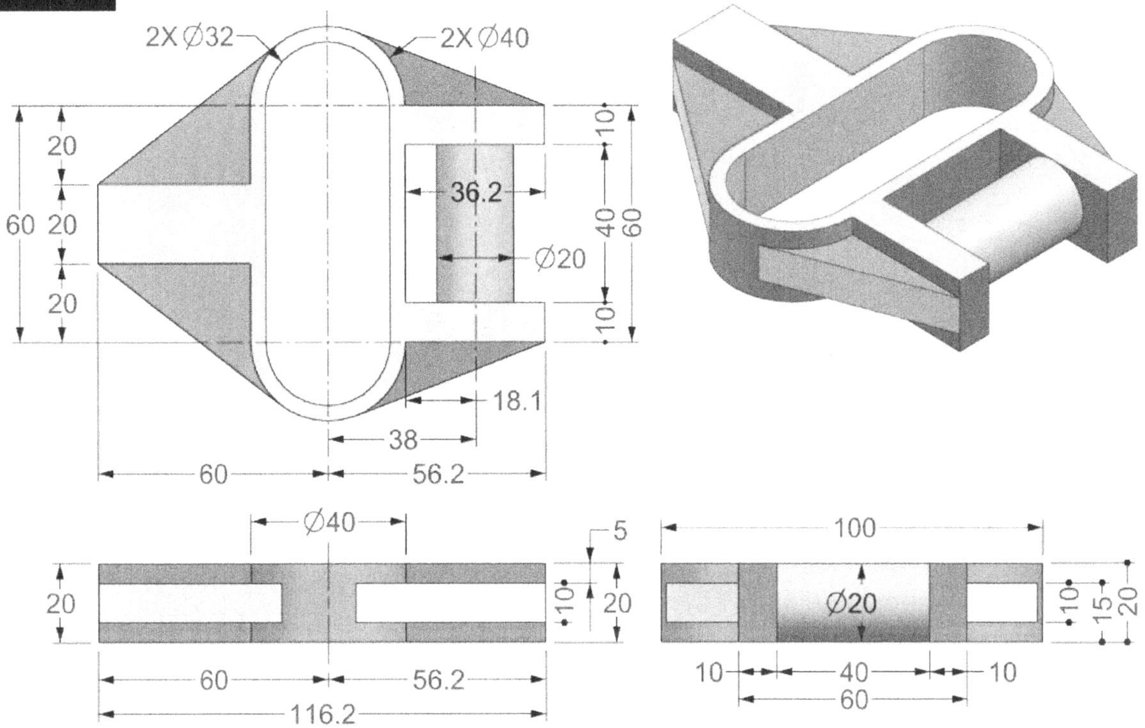

2X Ø32
2X Ø40
20
20
60
20
20
36.2
Ø20
10
40
60
10
18.1
38
60
56.2

Ø40
5
20
20
10
60
56.2
116.2

100
Ø20
10
15
20
10
40
10
60

EX-91

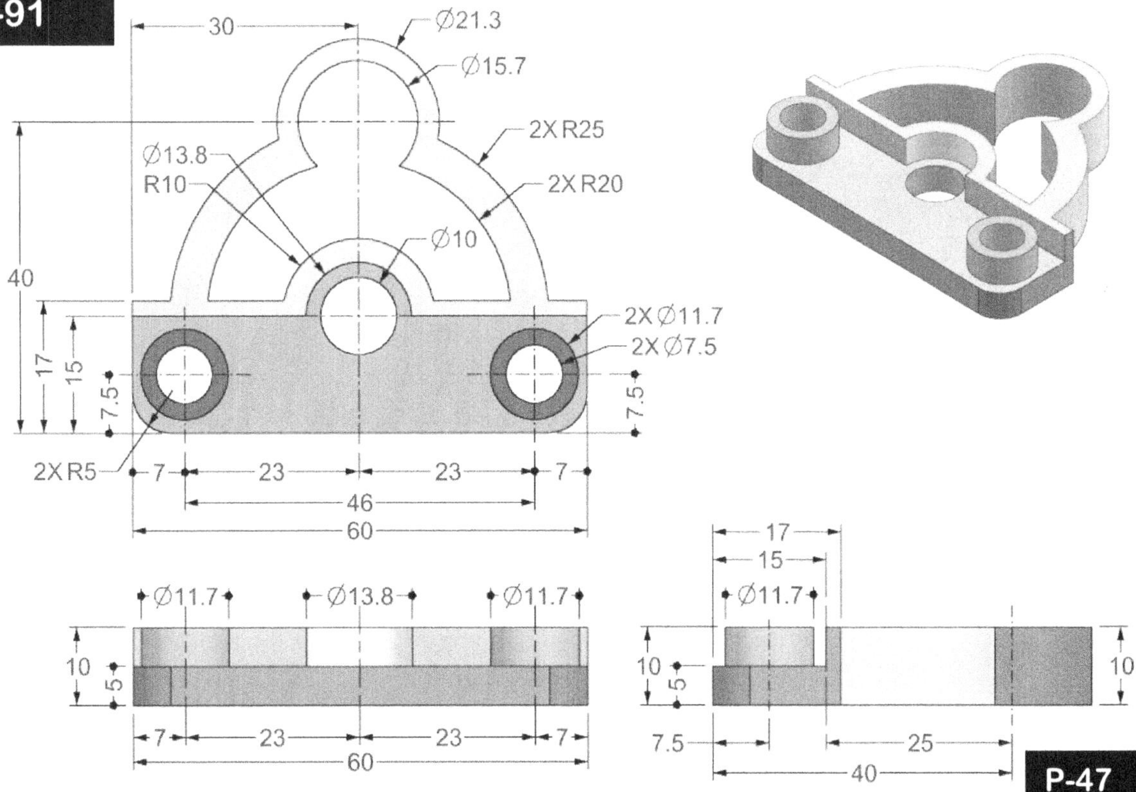

30
Ø21.3
Ø15.7
Ø13.8
R10
2X R25
2X R20
Ø10
40
17
15
7.5
2X Ø11.7
2X Ø7.5
7.5
2X R5
7
23
23
7
46
60

Ø11.7
Ø13.8
Ø11.7
10
5
7
23
23
7
60

17
15
Ø11.7
10
5
10
7.5
25
40

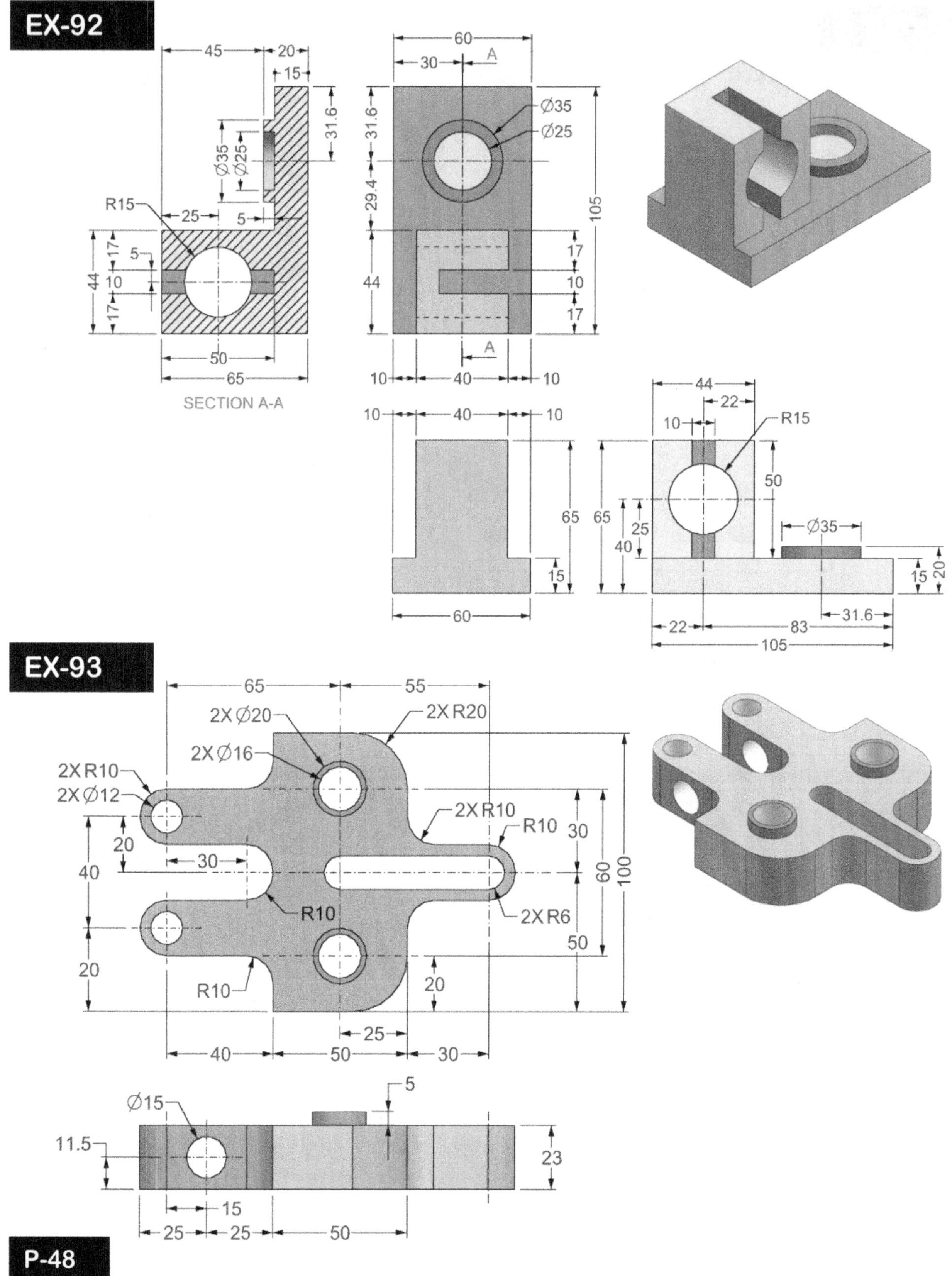

SECTION A-A
EX-93
P-48

EX-94
10
40
15
Ø12
SECTION A-A
A
R7.8
R10.7
2X R25
2X R20
R10
Ø12
25
2
17
R5
R5
15
30
30
A
EX-95
Ø60
Ø20
R2
R2
R2
R2
R2
10
40
60
70
100
Ø80
Ø60
Ø40
Ø20
R20
A
A
Ø10
100
Ø60
Ø40
Ø10
Ø20
R2
R2
R2
R2
R2
R2
R2
R2
R2
R2
10
40
60
70
10
100
SECTION A-A
P-49

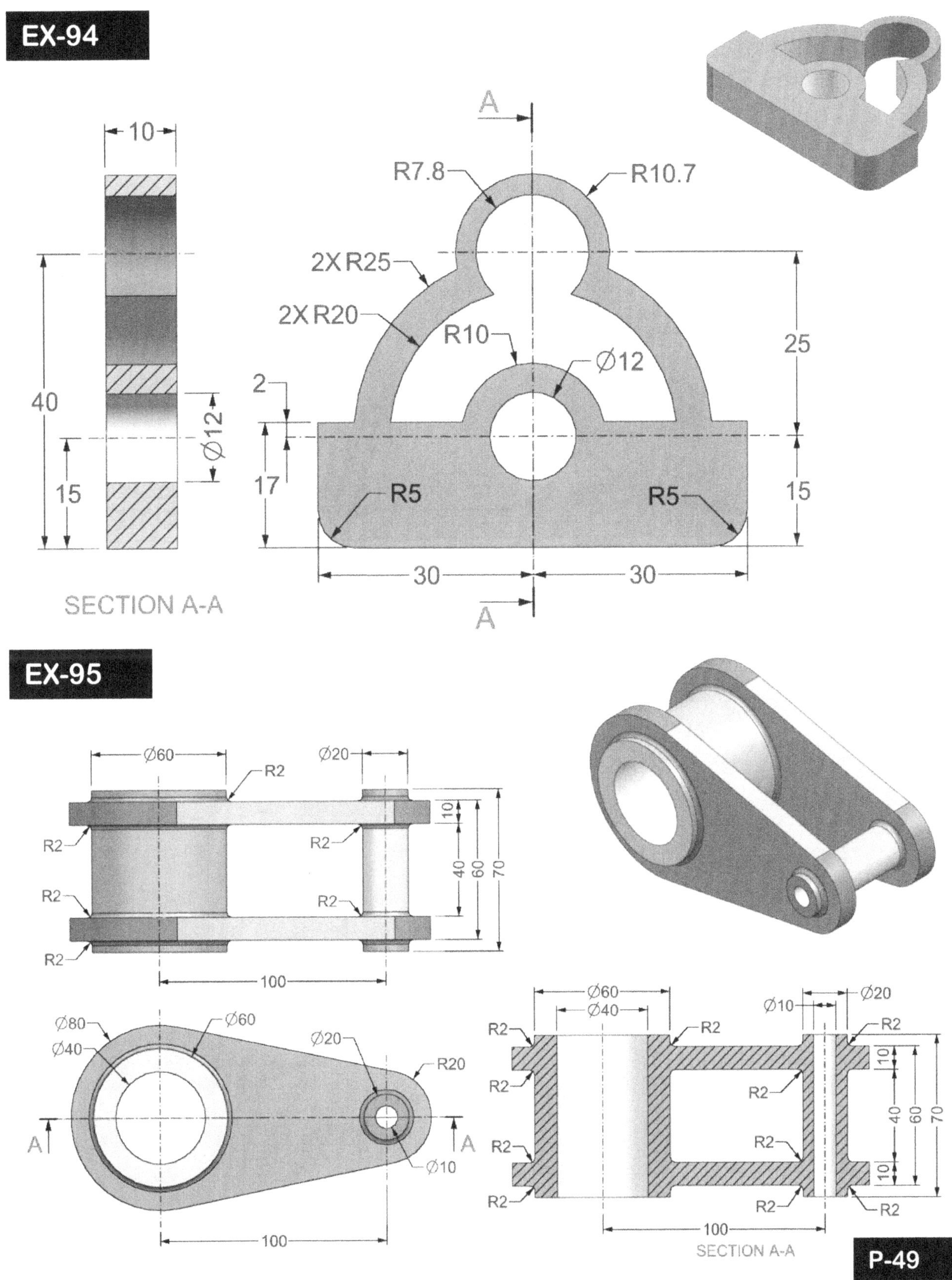

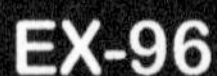

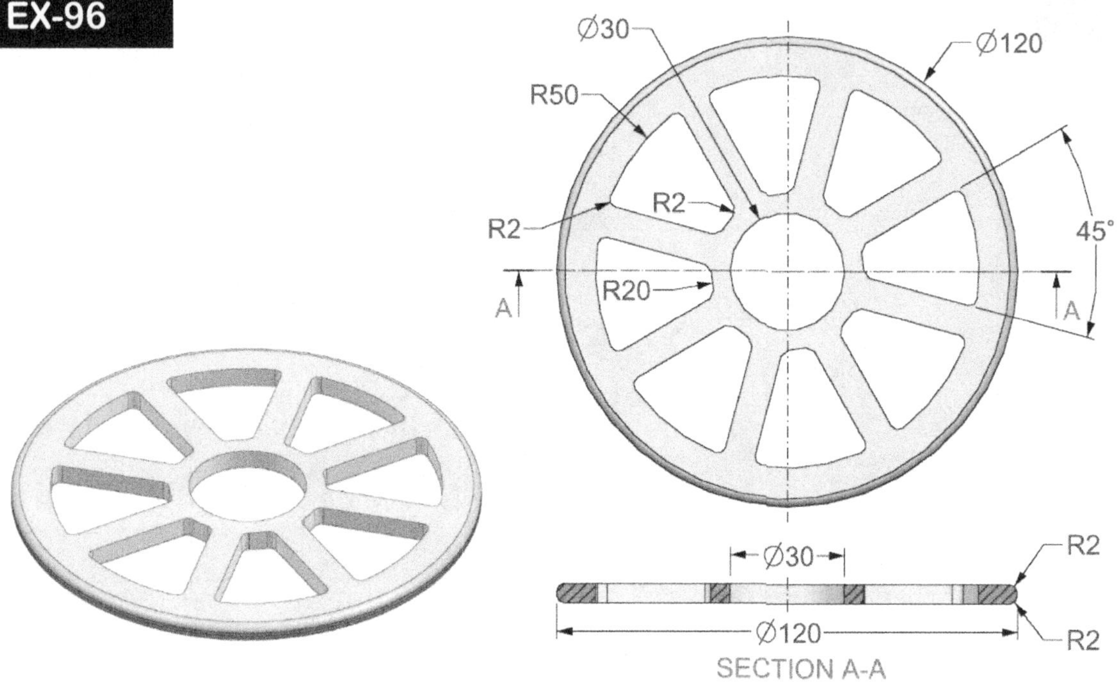

Ø30
Ø120
R50
45°
R2
R2
R2
R20
A
A
Ø30
Ø120
R2
R2
SECTION A-A

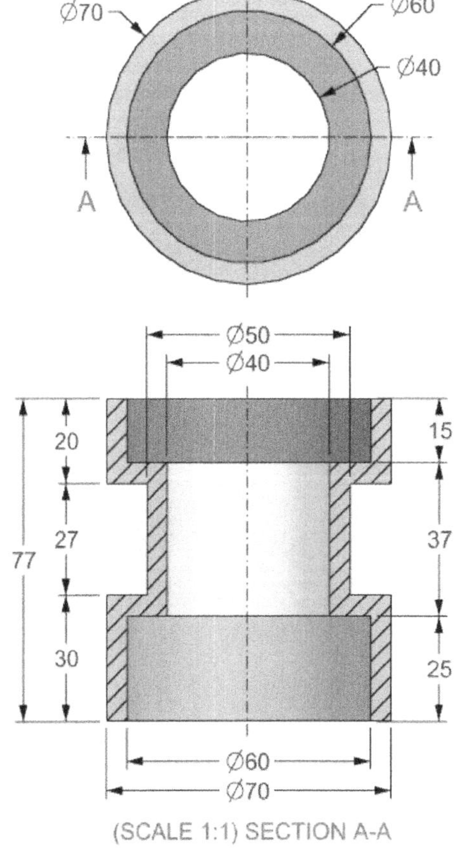

Ø70
Ø60
Ø40
A
A
Ø50
Ø40
20
15
27
37
77
30
25
Ø60
Ø70
(SCALE 1:1) SECTION A-A

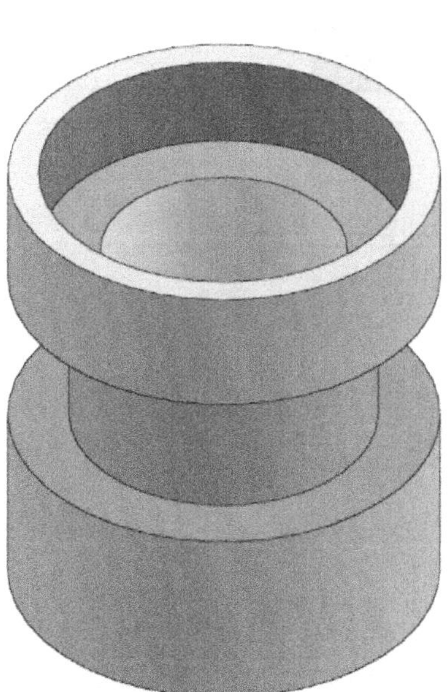

EX-98 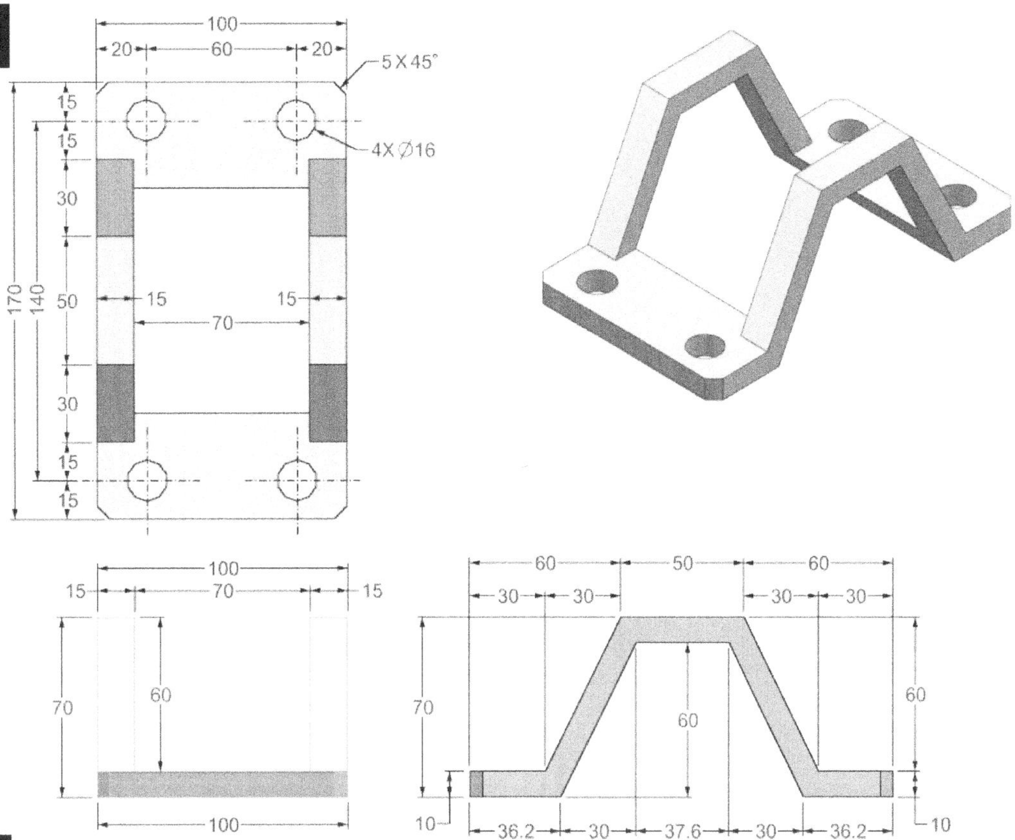
100
20
60
20
5 X 45°
15
15
4X Ø16
30
50
15
15
70
30
15
15
170
140
100
15
70
15
70
60
100
60
50
60
30
30
30
30
70
60
60
10
36.2
30
37.6
30
36.2
10
EX-99 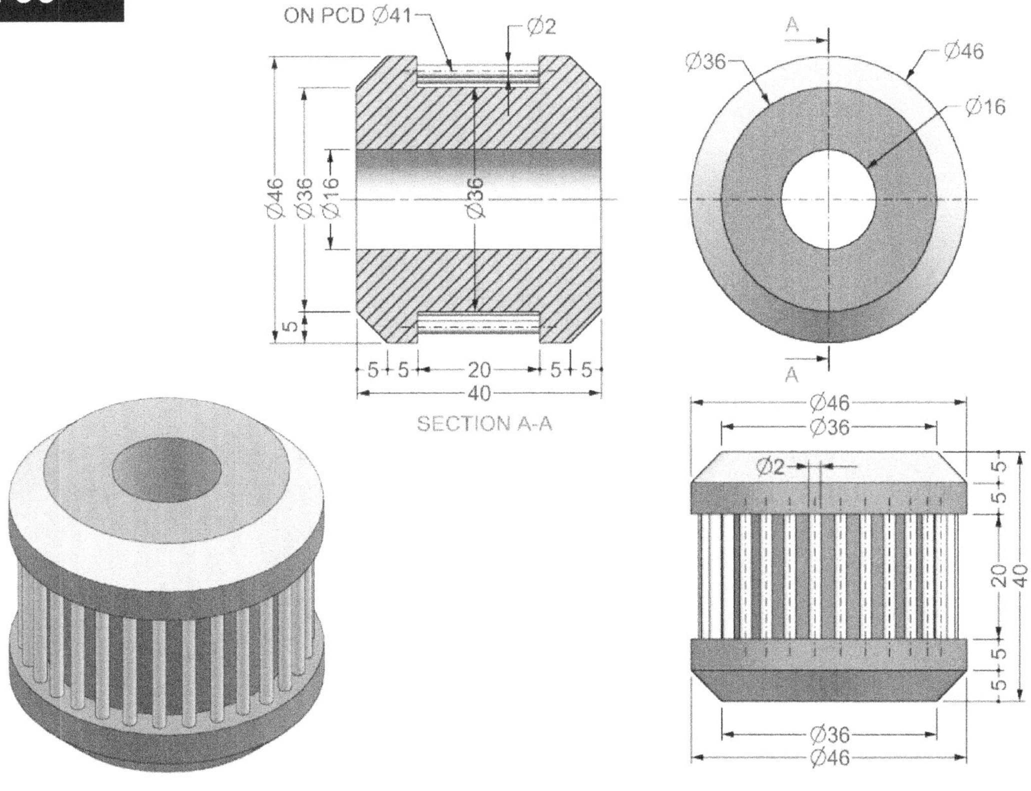
ON PCD Ø41
Ø2
Ø46
Ø36
Ø16
Ø36
A
Ø36
A
5
5
5
20
5
5
40
SECTION A-A
Ø46
Ø36
Ø2
5
5
5
5
20
40
Ø36
Ø46
P-51

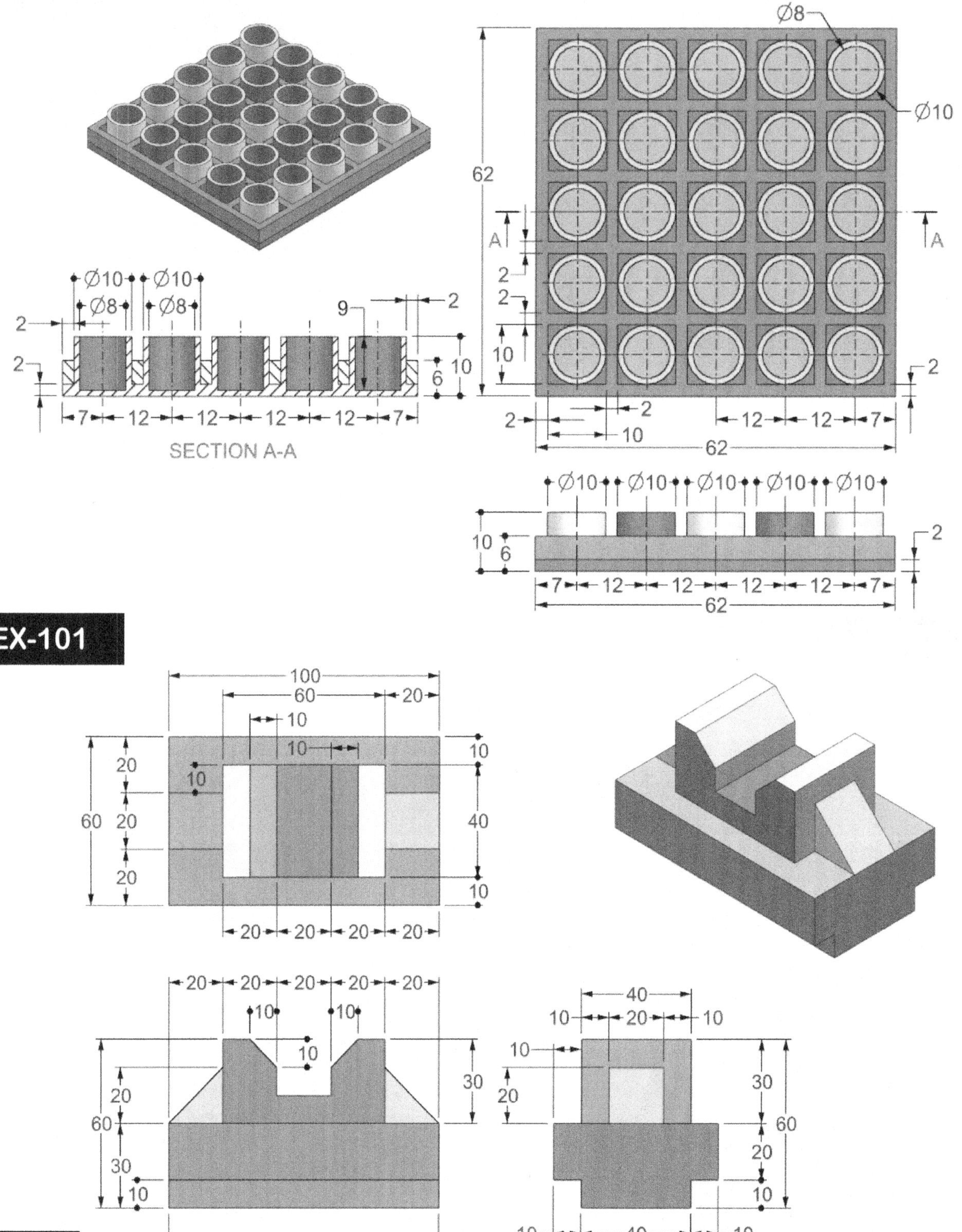

EX-102
4X ⌀20
4X ⌀16
124
96
14
14
8
14
124
96
8
8
8
⌀20
⌀20
7.5
8
23
8
14
96
14
124
⌀20
⌀20
8
7.5
14
96
14
124
EX-103
3
7
7
10
10
10
23
10
23
23
107
23
5
23
3
23
23
23
23
113
10
10
3
7
16
8
113
10
10
5
3
16
8
107
P-53

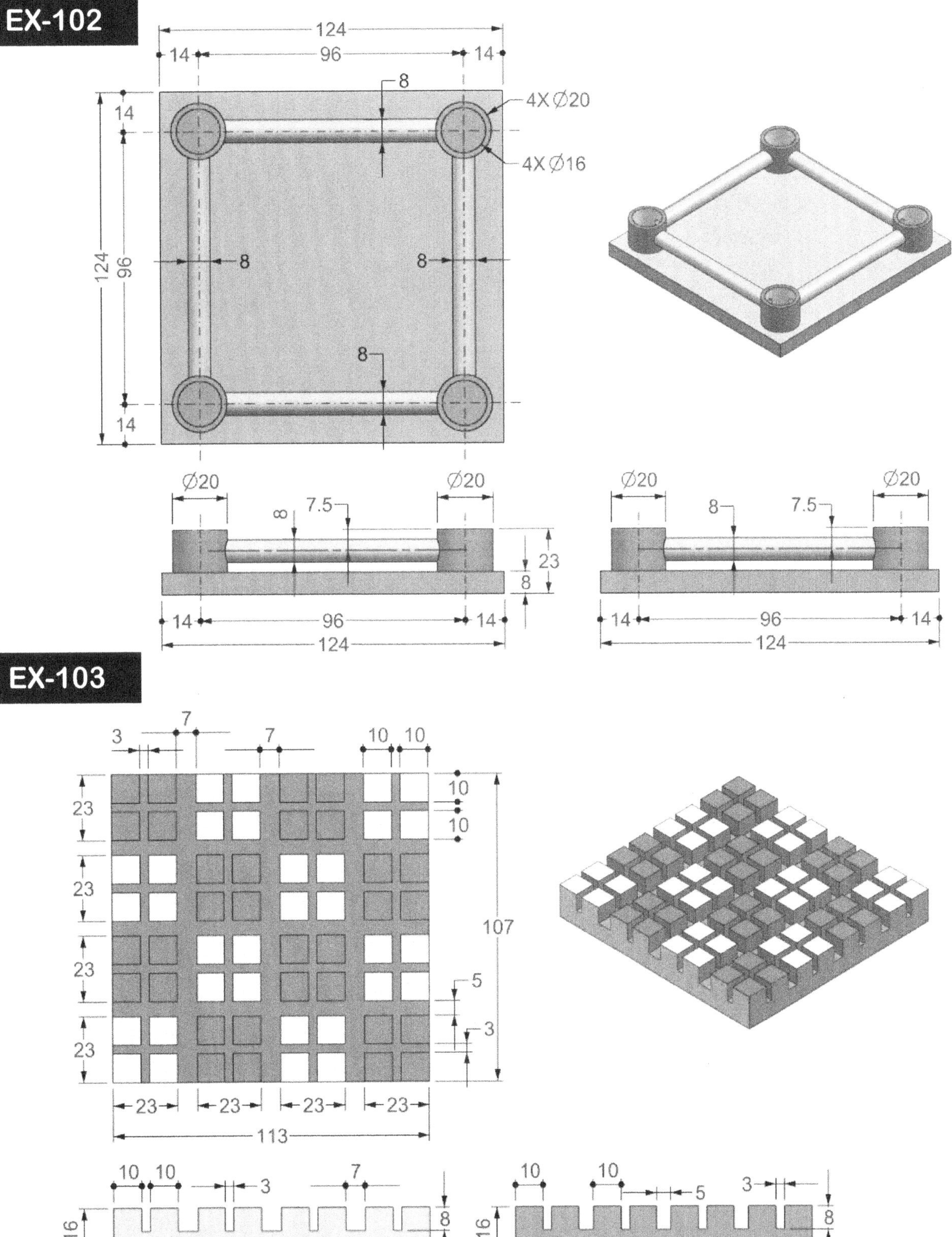

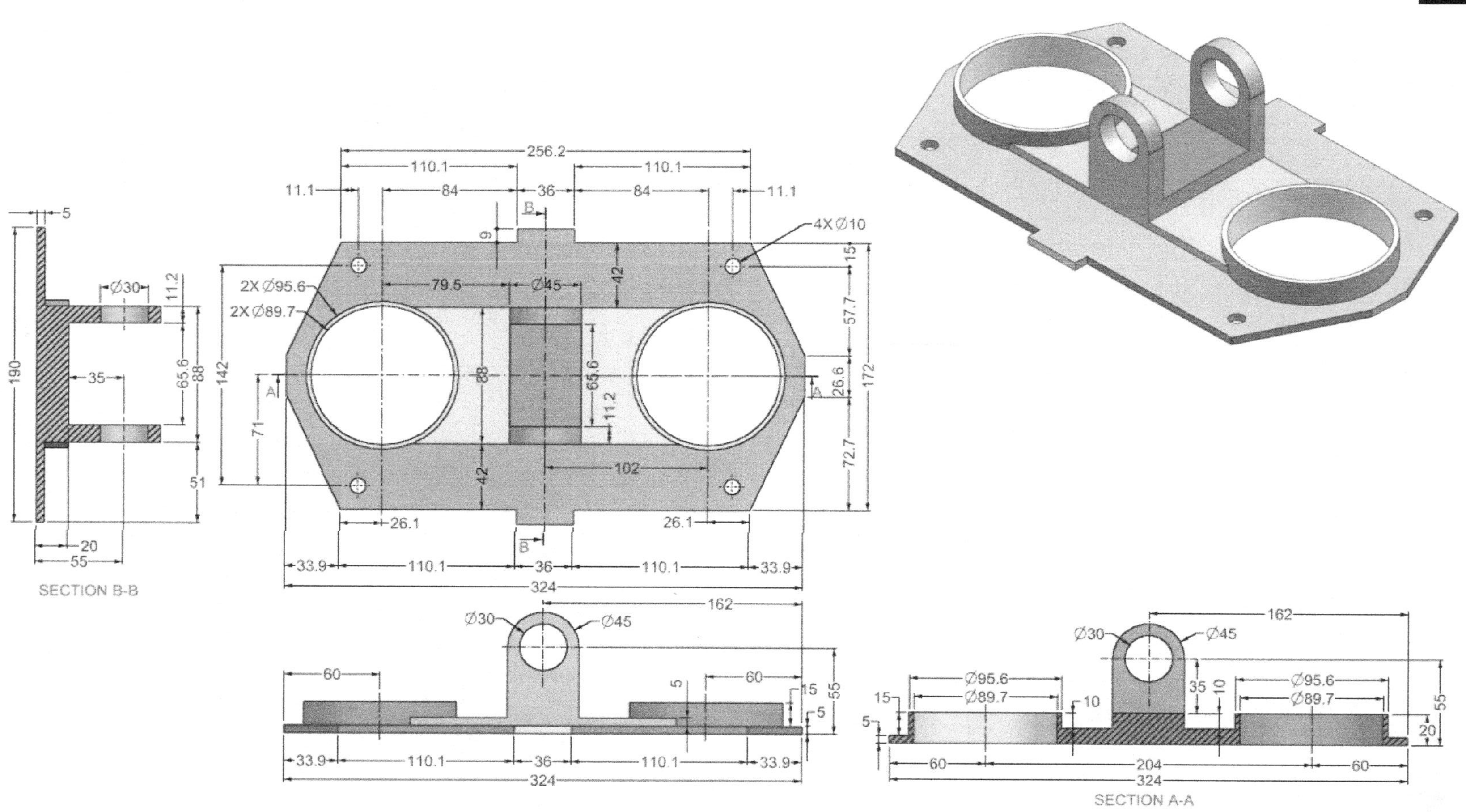

SECTION B-B
SECTION A-A
4X Ø10
2X Ø95.6
2X Ø89.7
Ø30
Ø45
Ø95.6
Ø89.7

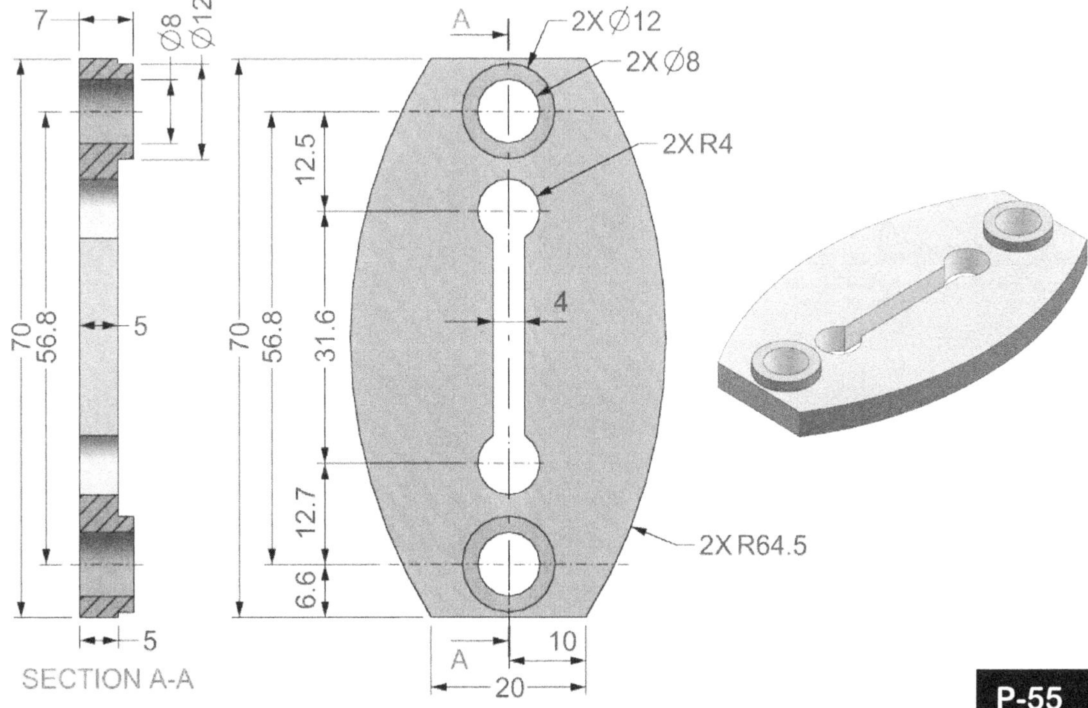
EX-105
324 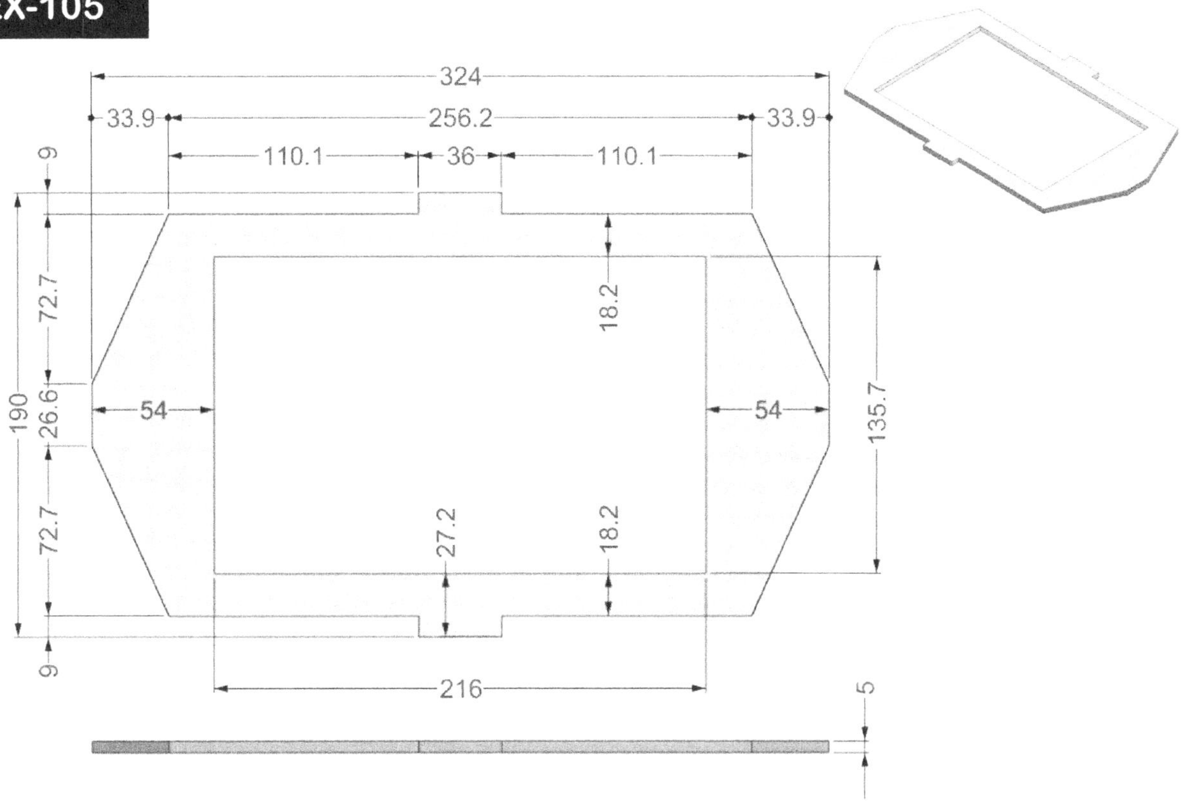
33.9
256.2
33.9
9
110.1
36
110.1
72.7
18.2
190
26.6
54
54
135.7
72.7
27.2
18.2
9
216
5
EX-106
7
Ø8
Ø12
A
2X Ø12
2X Ø8
12.5
2X R4
70
56.8
70
56.8
5
31.6
4
12.7
2X R64.5
6.6
5
10
SECTION A-A
A
20
P-55

EX-107

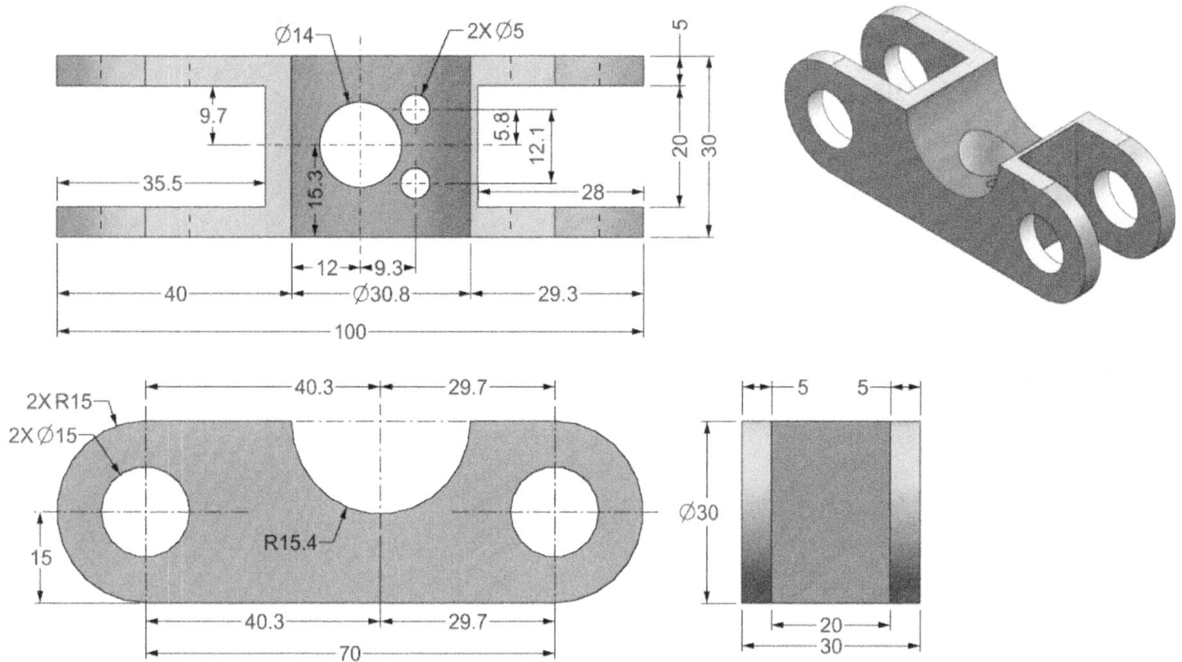

Ø14
2X Ø5
9.7
5.8
12.1
20
30
5
35.5
15.3
28
12
9.3
Ø30.8
40
29.3
100
2X R15
2X Ø15
40.3
29.7
5
5
R15.4
Ø30
15
40.3
29.7
70
20
30

EX-108

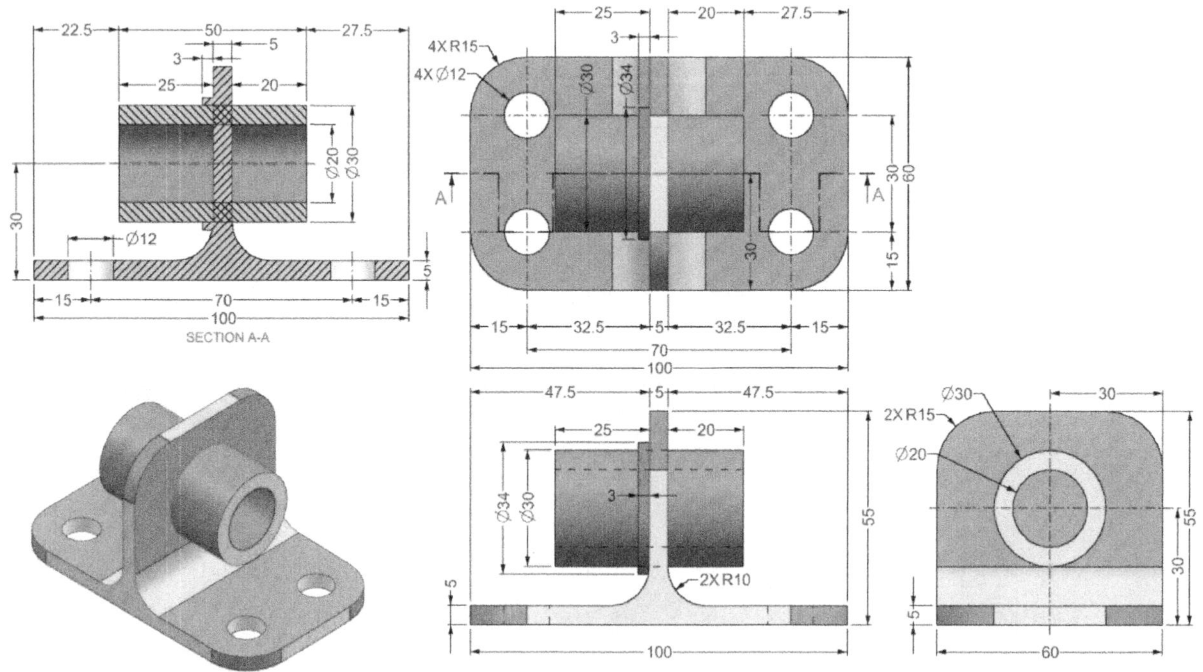

22.5
50
27.5
3
5
25
20
Ø20
Ø30
30
Ø12
15
70
15
100
SECTION A-A
25
20
27.5
3
4X R15
4X Ø12
Ø30
Ø34
A
A
30
60
30
15
15
32.5
5
32.5
15
70
100
47.5
5
47.5
25
20
Ø34
Ø30
3
5
55
2X R10
100
Ø30
2X R15
Ø20
30
55
30
5
60

P-56

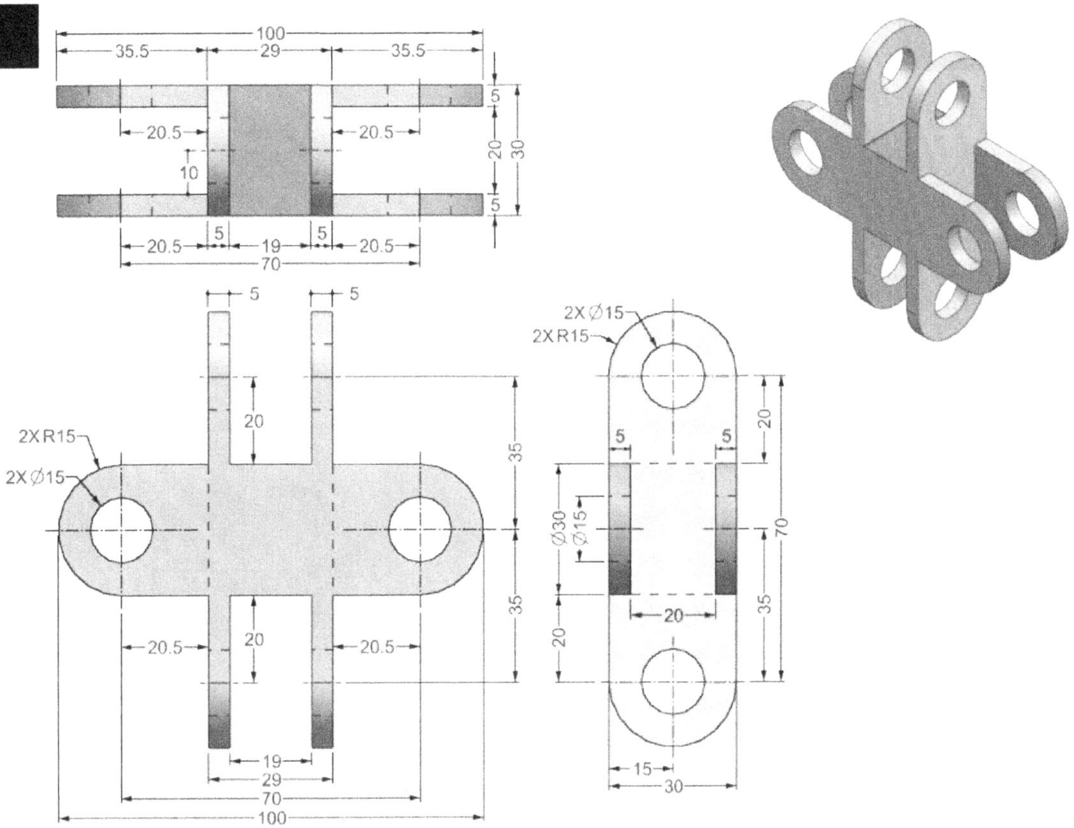

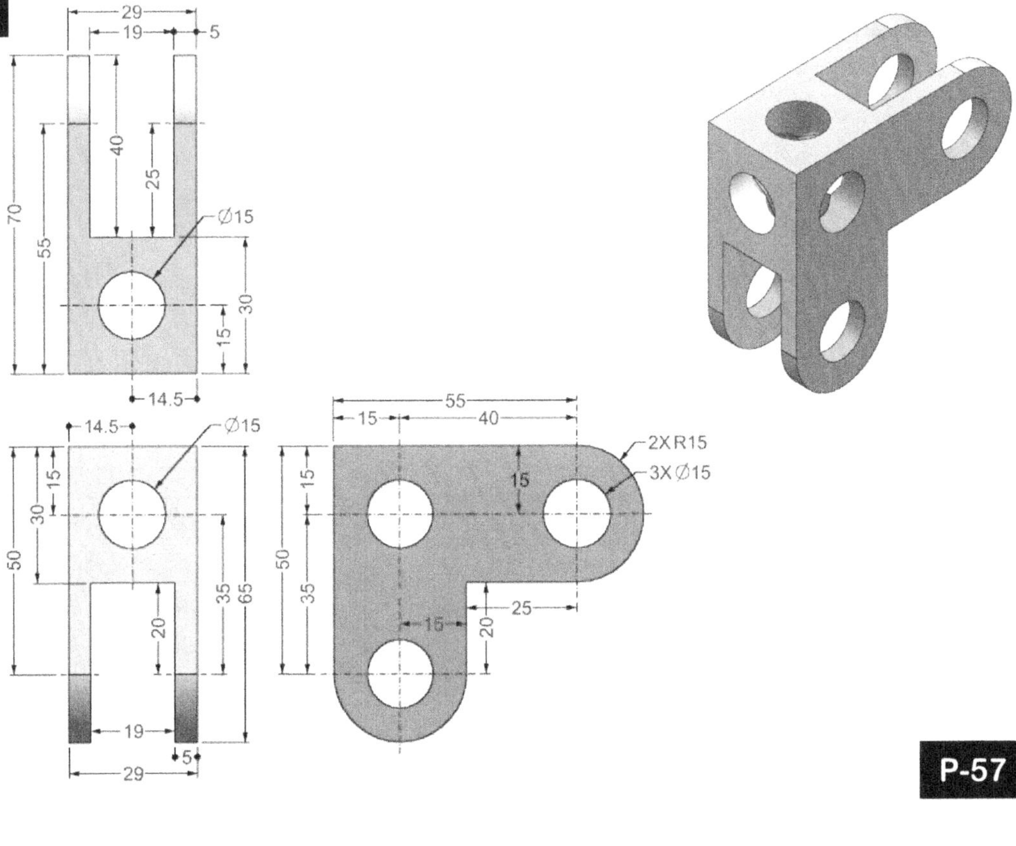

EX-111 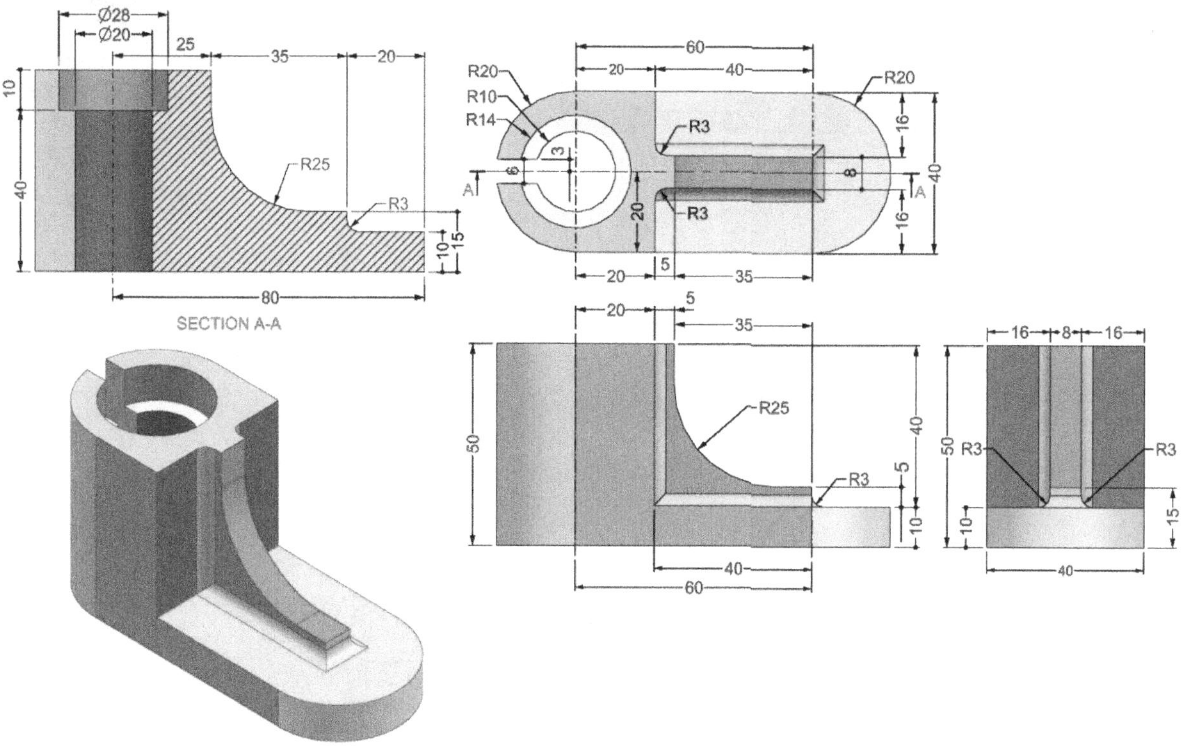
Ø28
Ø20
25
35
20
10
40
R25
R3
10
15
80
SECTION A-A
R20
R10
R14
60
20
40
R20
16
R3
3
6
A
40
8
A
20
R3
5
20
35
20
5
35
16
8
16
50
R25
40
5
R3
50
R3
R3
10
40
10
15
60
40
EX-112 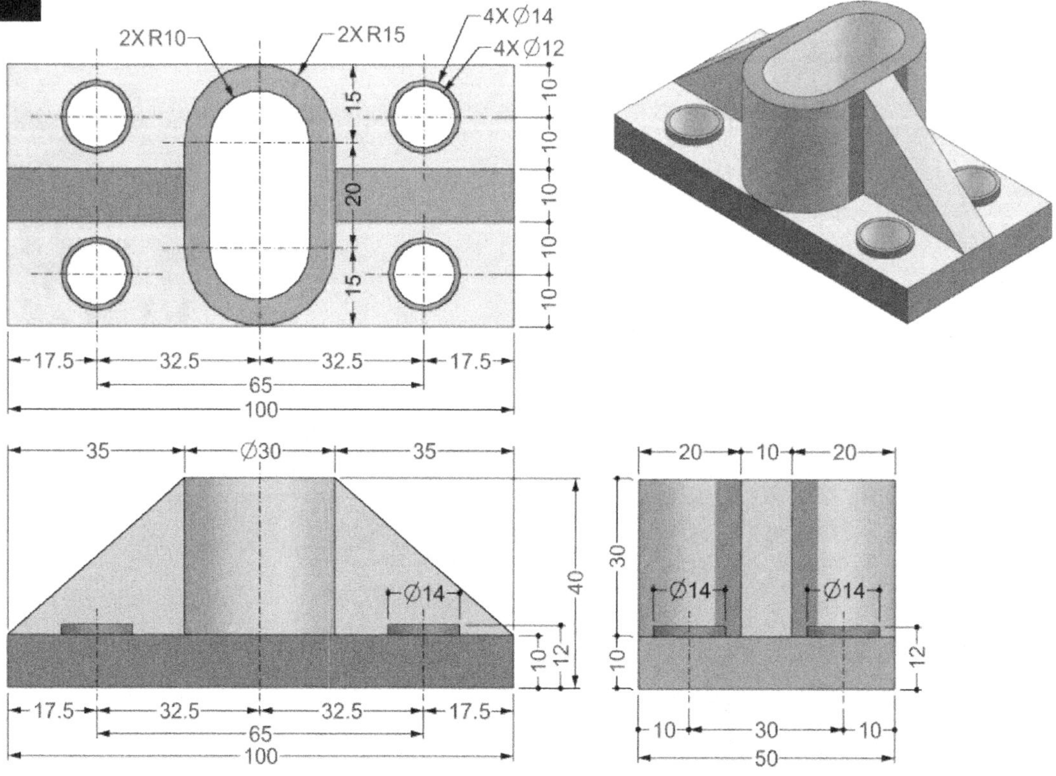
2X R10
2X R15
4X Ø14
4X Ø12
15
10
10
10
20
10
10
15
10
17.5
32.5
32.5
17.5
65
100
35
Ø30
35
Ø14
40
10
12
17.5
32.5
32.5
17.5
65
100
20
10
20
30
Ø14
Ø14
10
12
10
30
10
50
P-58

EX-113

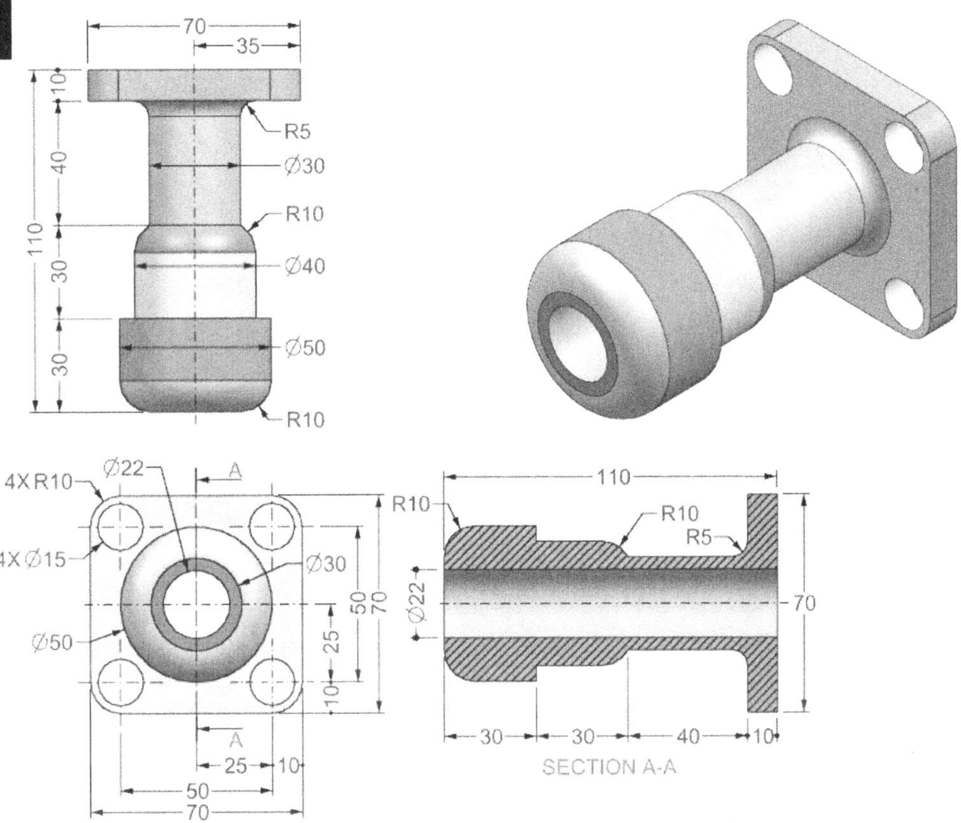

70
35
10
40
110
30
30
R5
Ø30
R10
Ø40
Ø50
R10

4X R10
Ø22
A
4X Ø15
Ø30
Ø50
50
70
25
10
A
25
10
50
70

110
R10
R10
R5
Ø22
70
30
30
40
10
SECTION A-A

EX-114

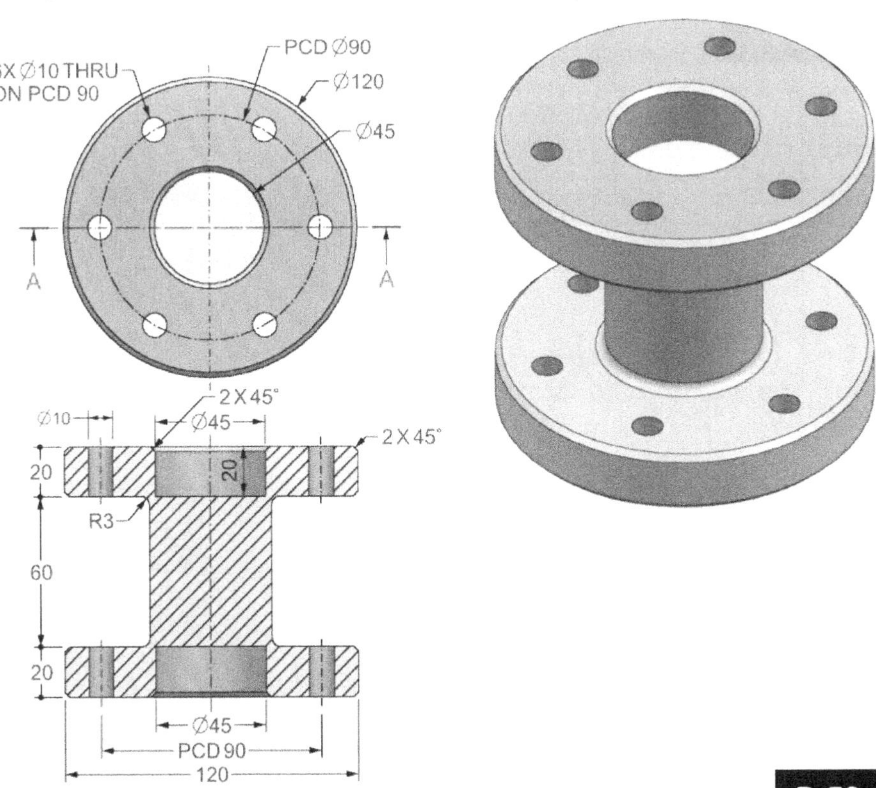

6X Ø10 THRU
ON PCD 90
PCD Ø90
Ø120
Ø45
A
A

Ø10
2 X 45°
Ø45
2 X 45°
20
20
R3
60
20
Ø45
PCD 90
120
SECTION A-A

EX-115

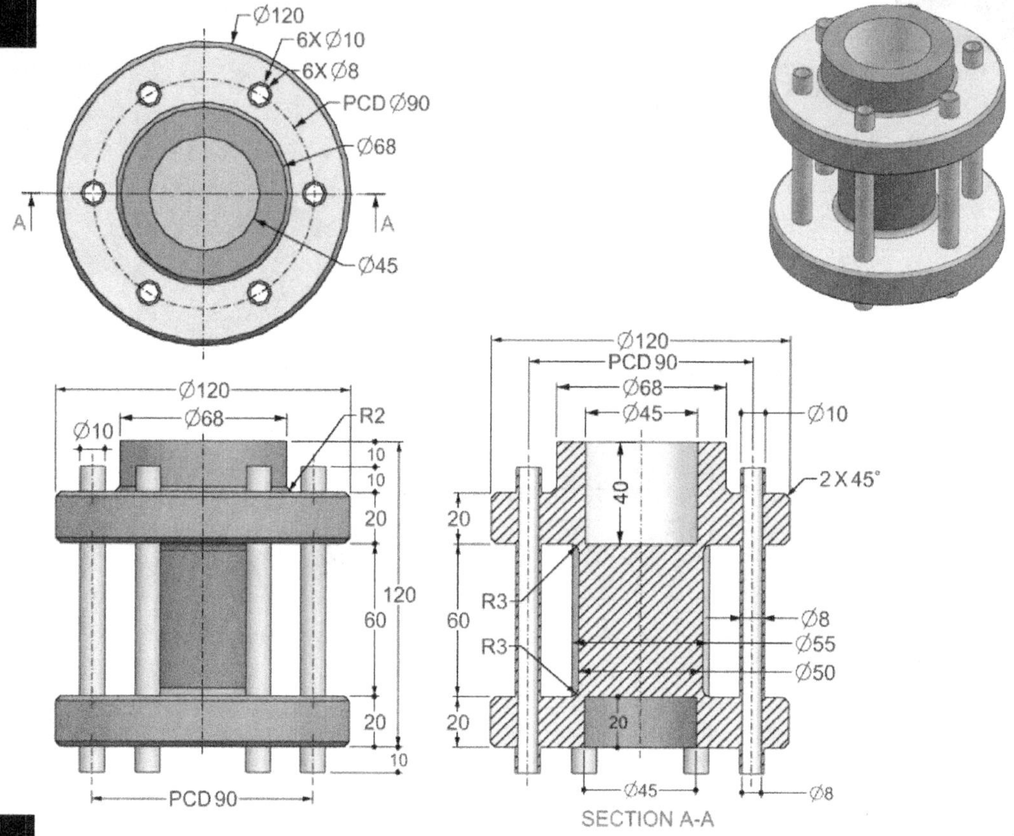

EX-116

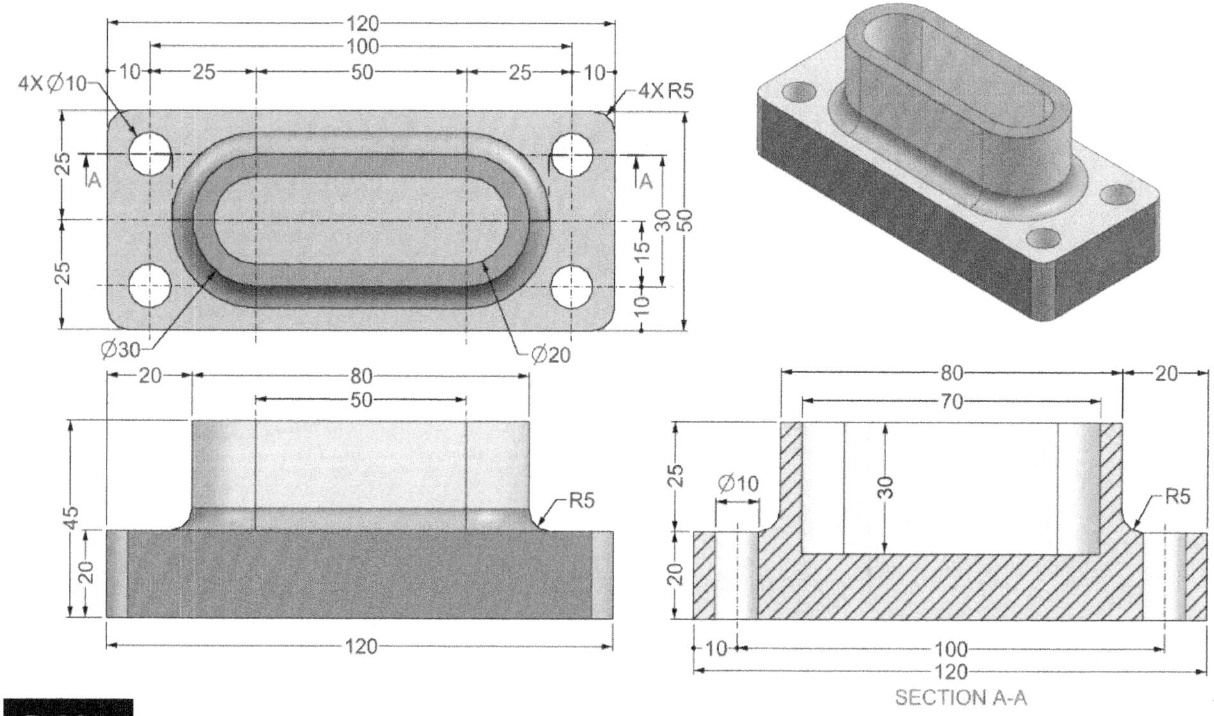

P-60

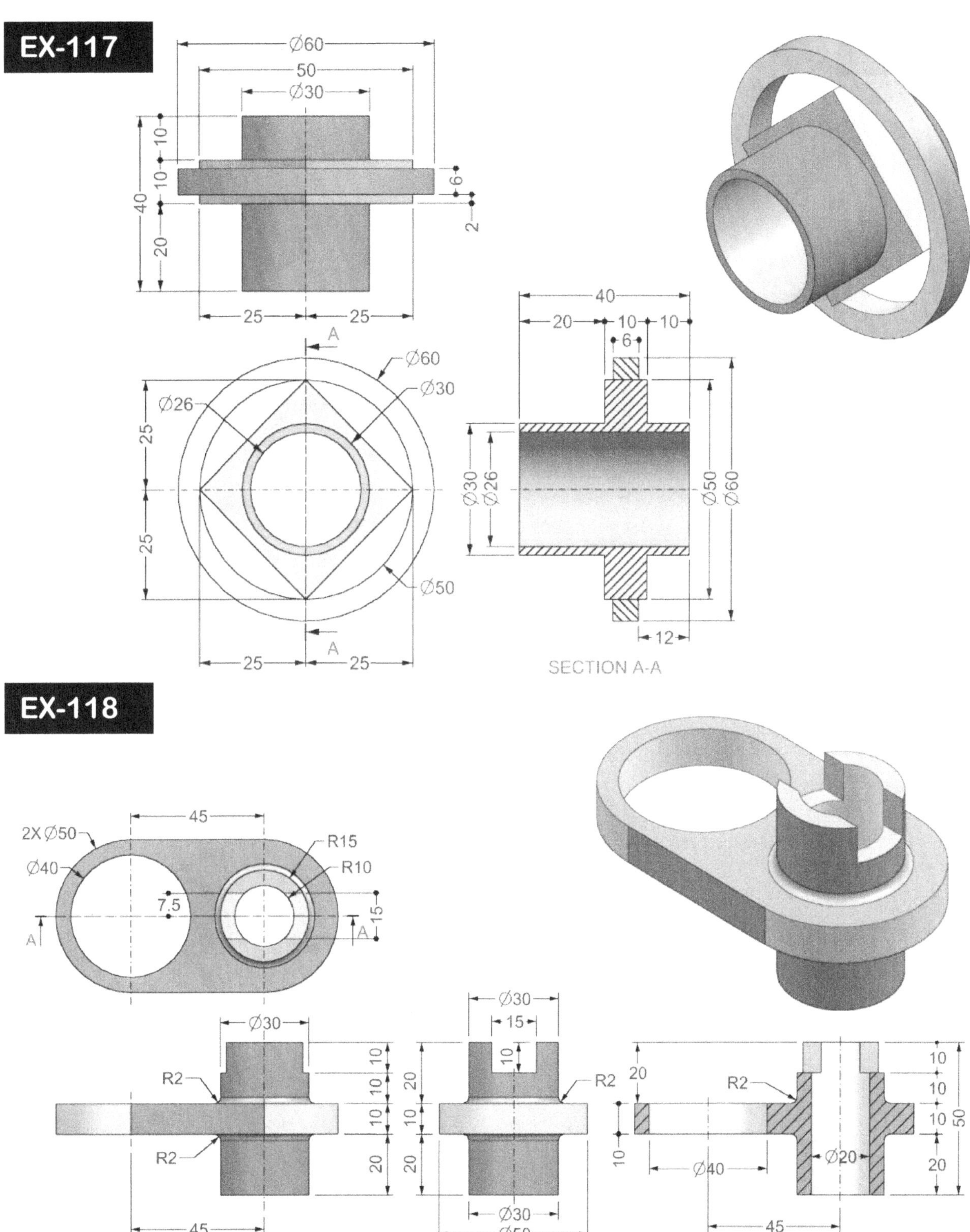

EX-117
Ø60
50
Ø30
10
10
40
6
20
2
25
25
A
A
Ø60
Ø30
Ø26
25
25
Ø50
25
25
40
20
10
10
6
Ø30
Ø26
Ø50
Ø60
12
SECTION A-A
EX-118
45
2X Ø50
Ø40
R15
R10
7.5
A
A
15
Ø30
Ø30
15
Ø30
R2
10
10
10
20
10
10
20
10
10
20
20
20
20
R2
20
R2
20
10
45
Ø30
Ø40
Ø20
Ø50
45
50
SECTION A-A
P-61

EX-119
Ø190 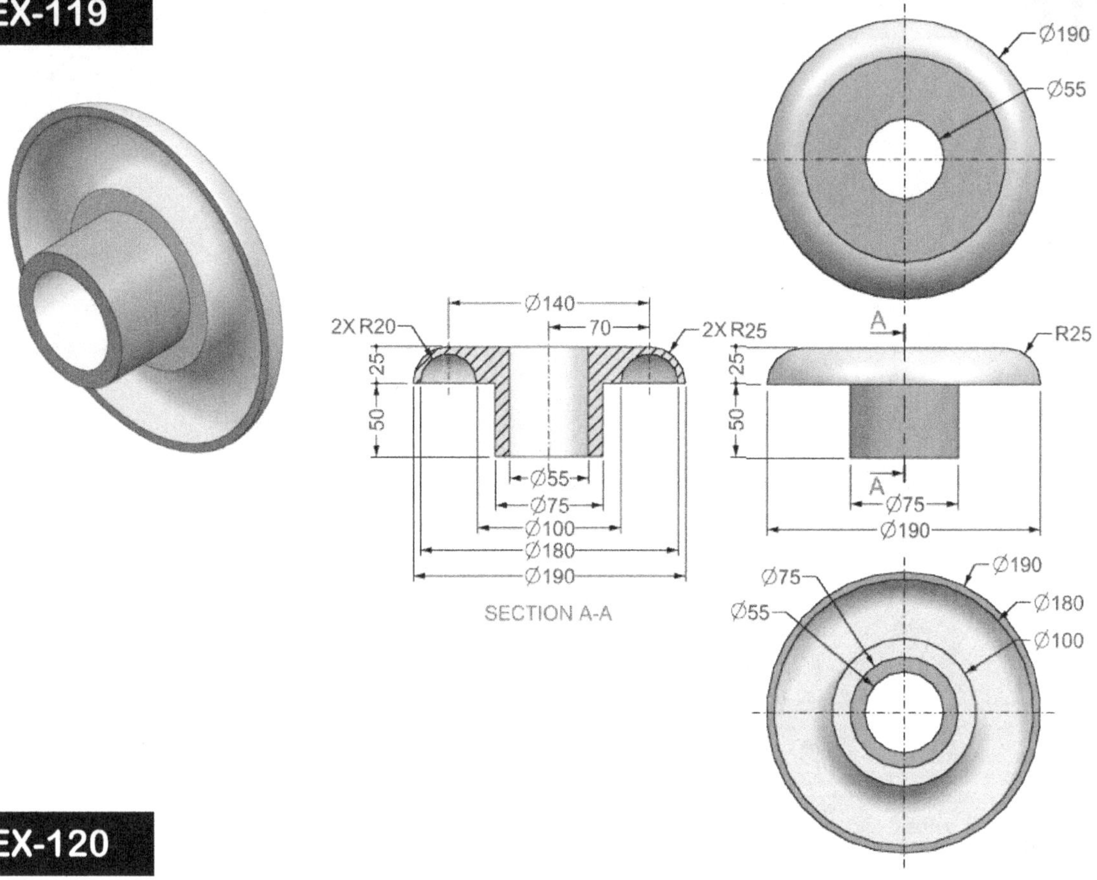
Ø55
Ø140
70
2X R20
2X R25
25
50
Ø55
Ø75
Ø100
Ø180
Ø190
SECTION A-A
A
R25
25
50
A
Ø75
Ø190
Ø75
Ø55
Ø190
Ø180
Ø100
EX-120
Ø50 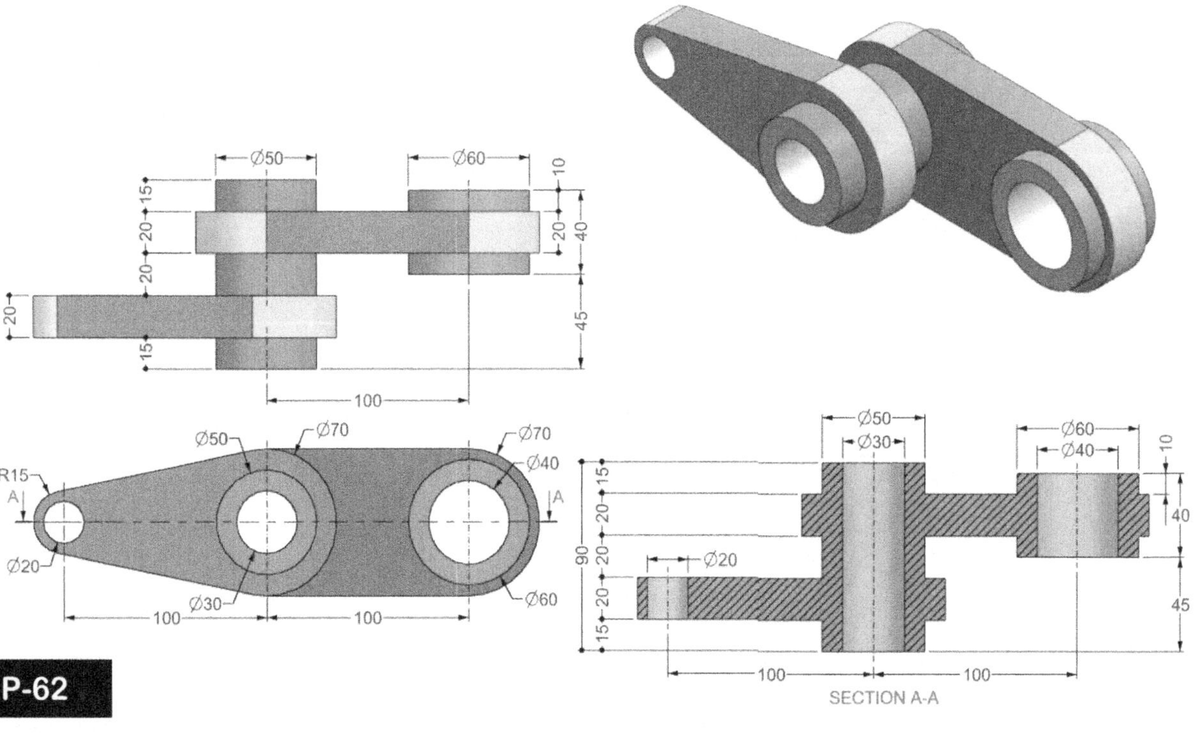
Ø60
10
15
20
20
20
20
15
20
40
45
100
Ø50
Ø70
Ø70
Ø40
R15
A
Ø20
Ø30
Ø60
100
100
A
Ø50
Ø30
Ø60
Ø40
10
15
20
20
20
Ø20
90
40
15
20
45
100
100
SECTION A-A
P-62

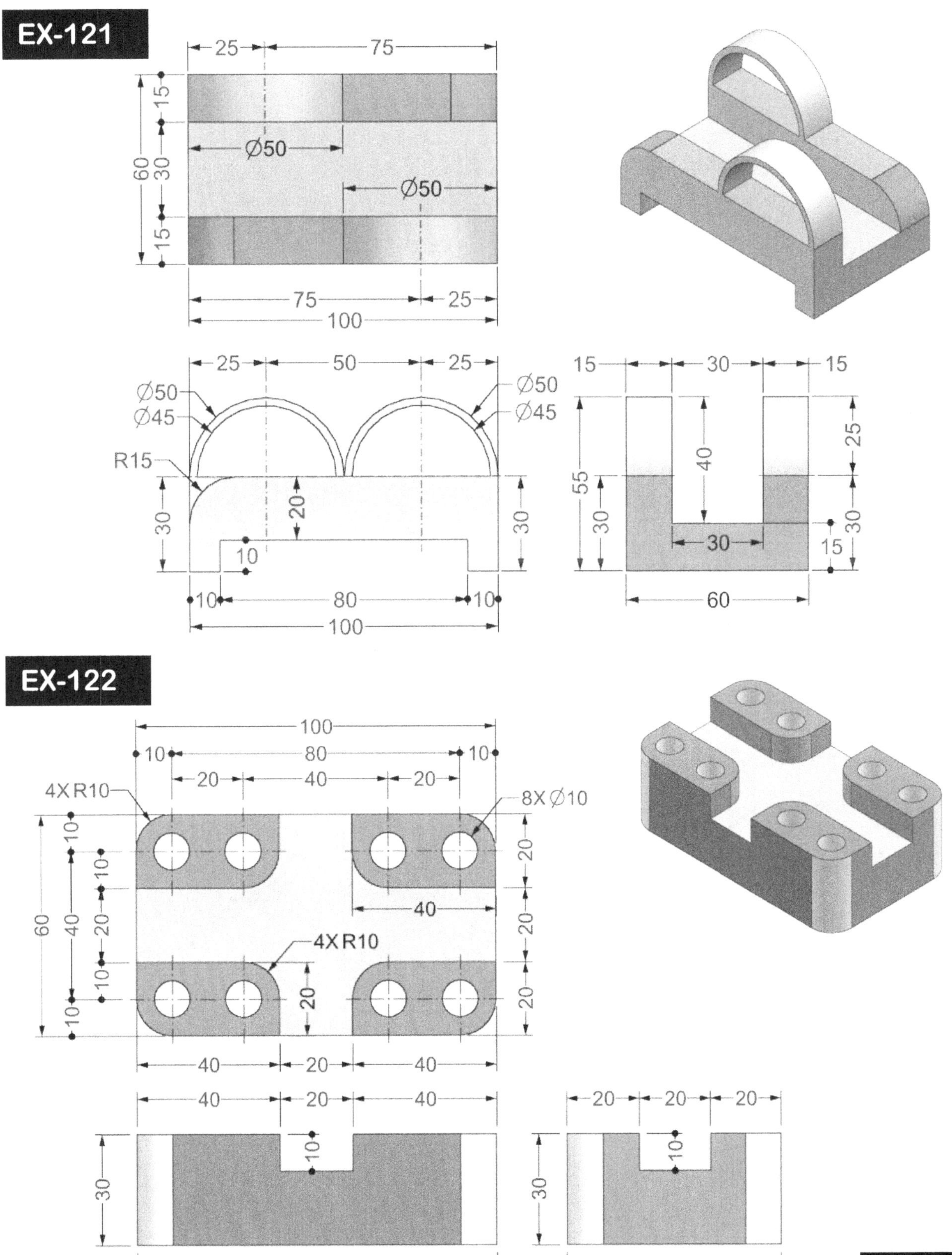

EX-121
25
75
15
60
30
15
Ø50
Ø50
75
25
100
25
50
25
Ø50
Ø45
Ø50
Ø45
R15
30
20
30
10
10
80
10
100
15
30
15
25
40
55
30
30
30
15
60
EX-122
100
10
80
10
20
40
20
4X R10
8X Ø10
10
10
10
20
10
10
40
40
20
20
60
40
40
4X R10
20
40
20
40
40
20
40
10
30
20
20
20
10
30
100
60
P-63

EX-123

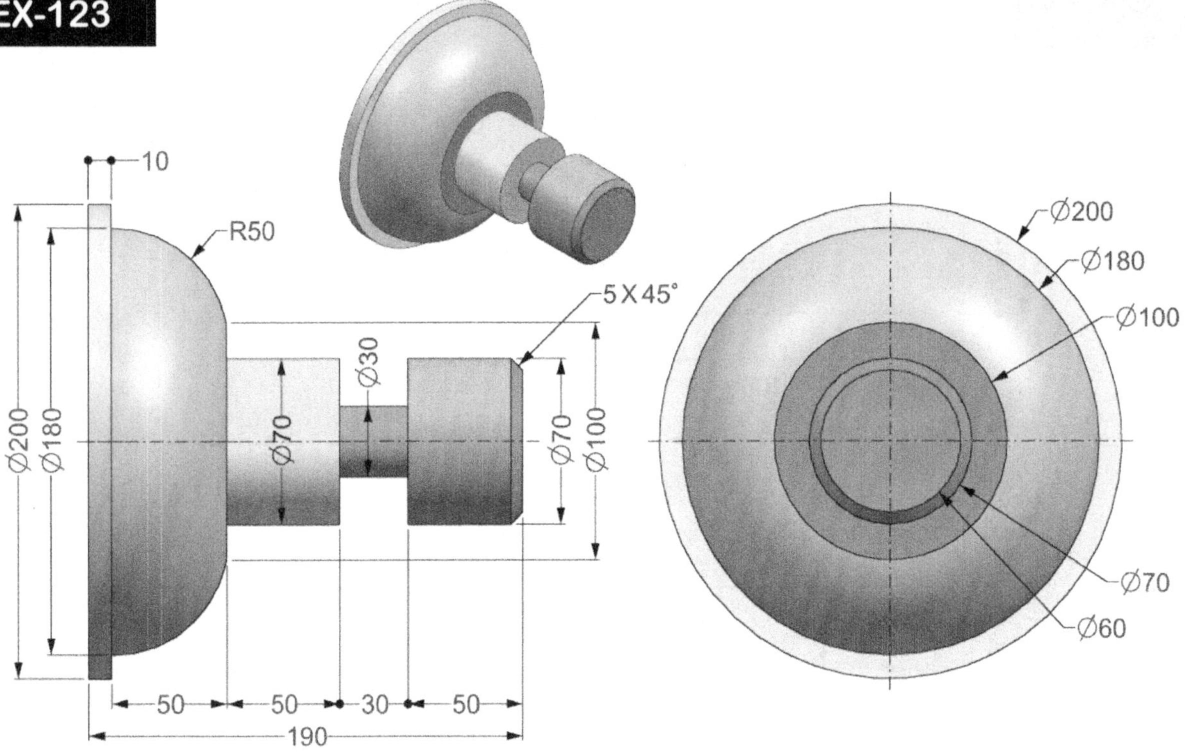

10
R50
Ø200
Ø180
Ø30
Ø70
5 X 45°
Ø70
Ø100
50 50 30 50
190

Ø200
Ø180
Ø100
Ø70
Ø60

EX-124

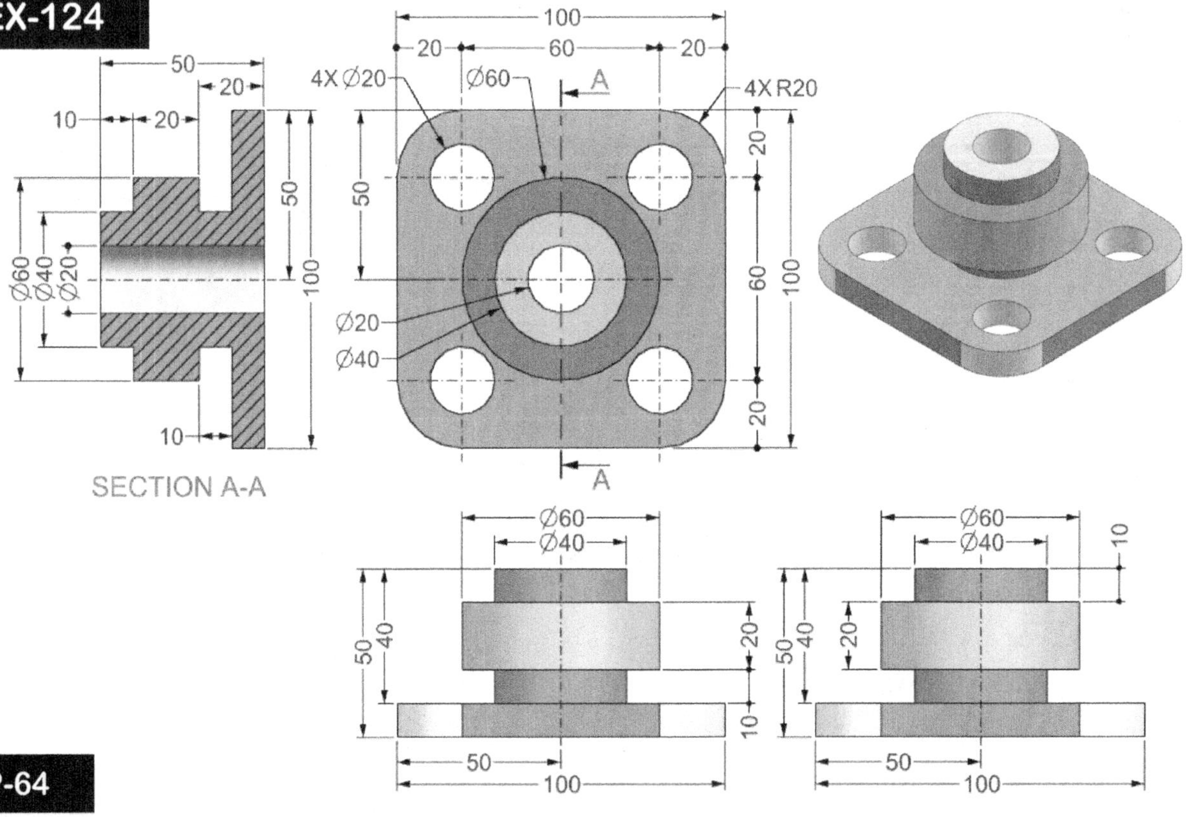

50
20
10
20
4X Ø20
Ø60
A
4X R20
50
Ø60
Ø40
Ø20
20
60
100
20
Ø20
Ø40
A

SECTION A-A

10

Ø60
Ø40
50
40
20
10
50
100

Ø60
Ø40
10
50
40
20
50
100

P-64

EX-125

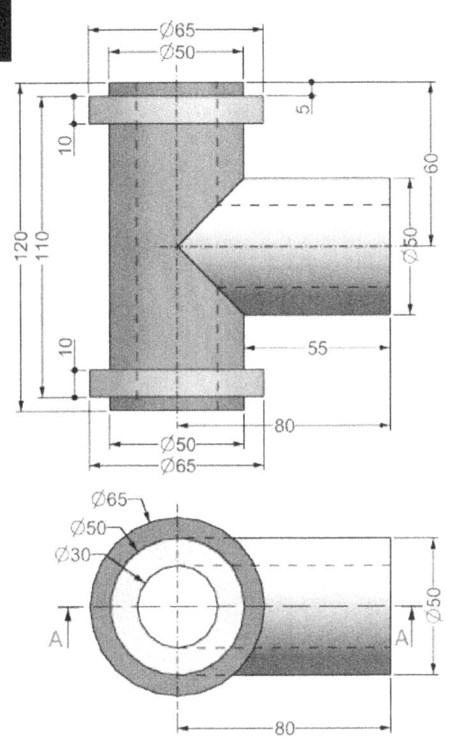
Ø65
Ø50
10
5
120
110
60
Ø50
10
55
Ø50
80
Ø65
Ø65
Ø50
Ø30
A
A
Ø50
80

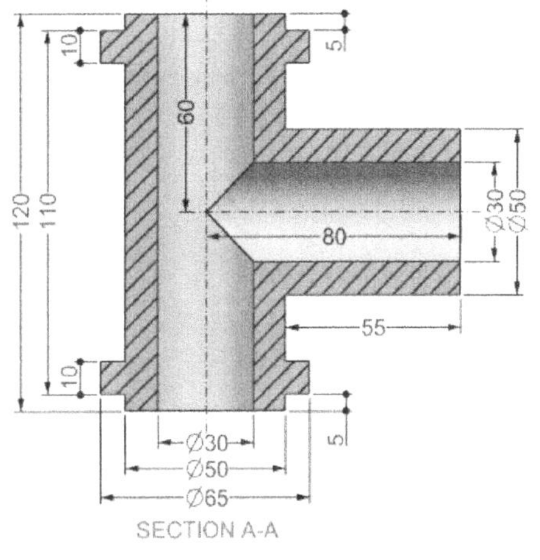
10
5
60
120
110
Ø30
Ø50
80
55
Ø30
Ø50
Ø65
10
5
SECTION A-A

EX-126

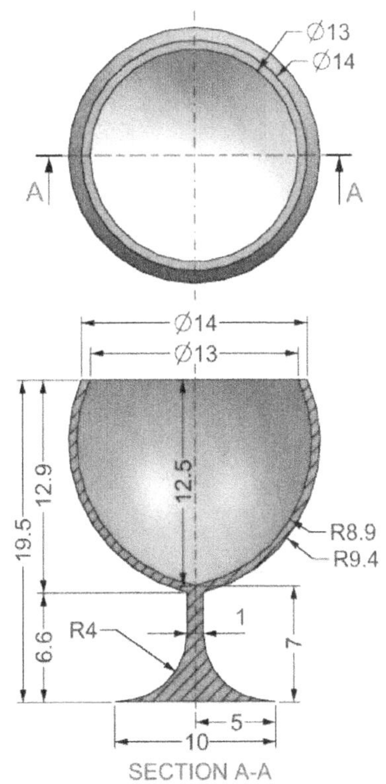
Ø13
Ø14
A
A
Ø14
Ø13
12.9
12.5
19.5
R8.9
R9.4
6.6
R4
1
7
5
10
SECTION A-A

P-65

EX-127

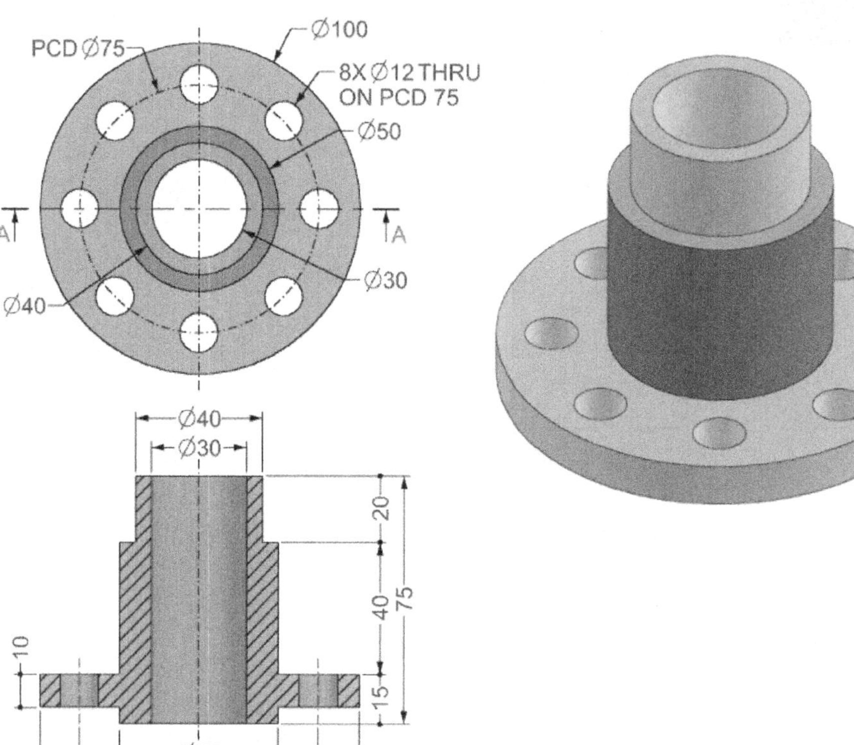

EX-128

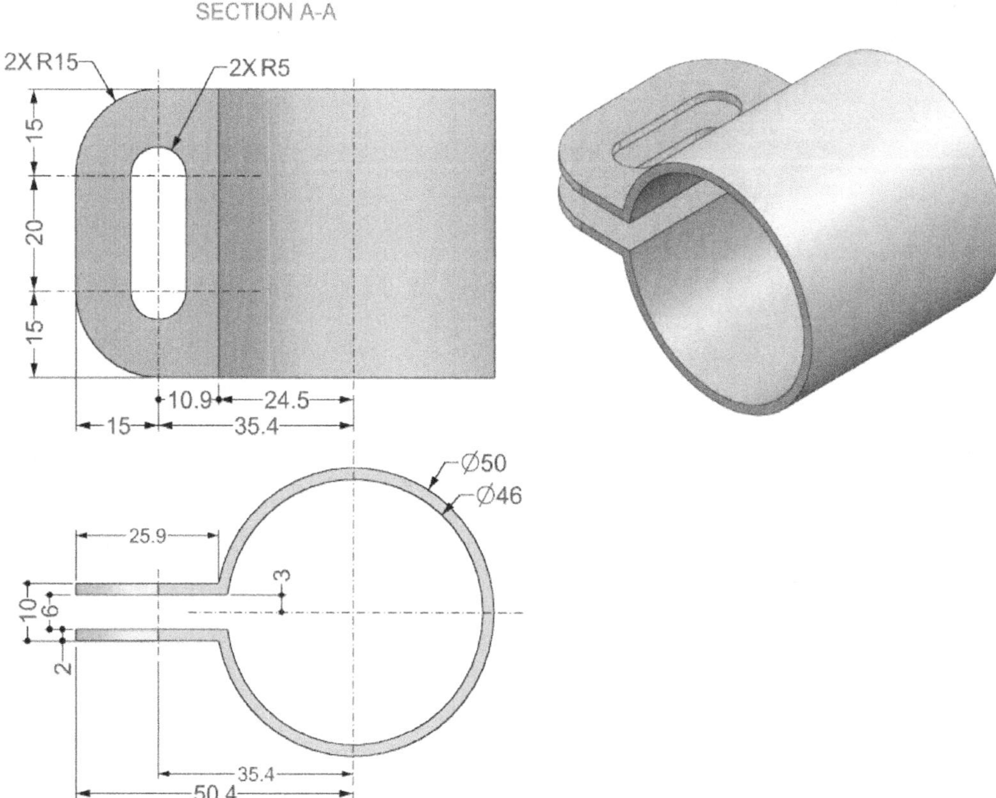

P-66

3X R20
3X Ø20
120°
R50
A
Ø80
A
PCD Ø140
Ø80
Ø20
15
Ø100
70
50
80
Ø20
SECTION A-A

EX-130

PCD Ø55
Ø70
A
A
8X Ø8
ON PCD 55
Ø30
Ø40
Ø40
45°
30
Ø8
Ø40
80
Ø8
35
15
10
Ø30
Ø40
PCD 55
Ø70
SECTION A-A

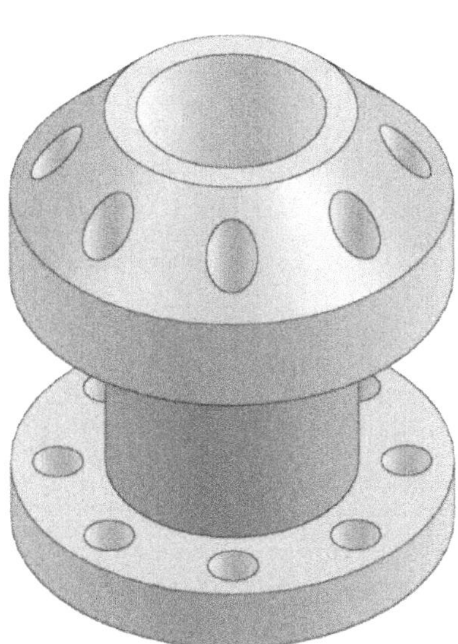

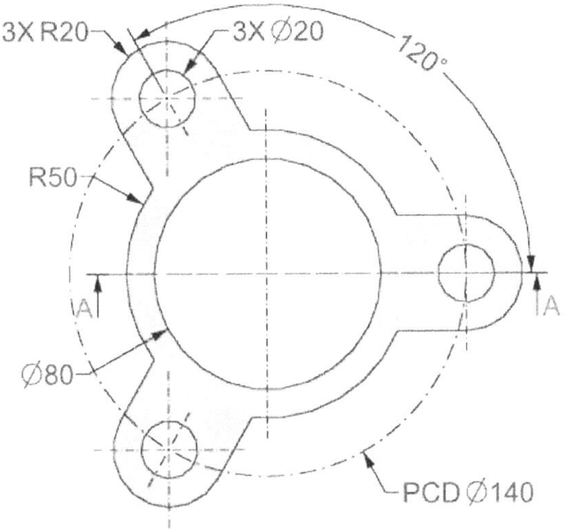

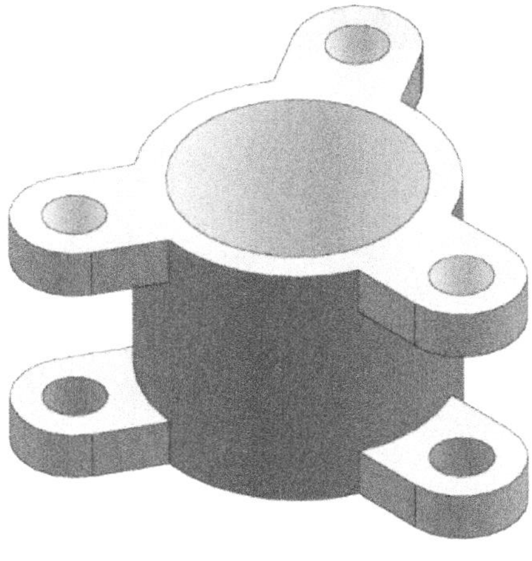

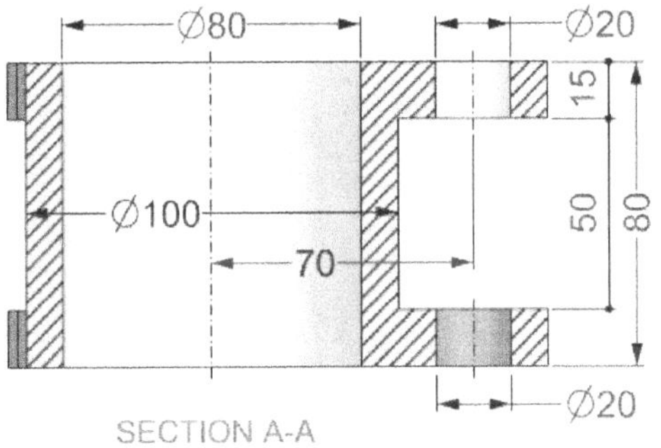

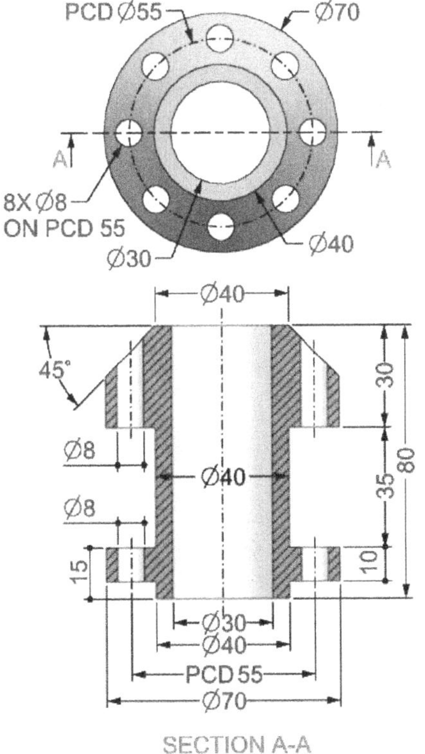

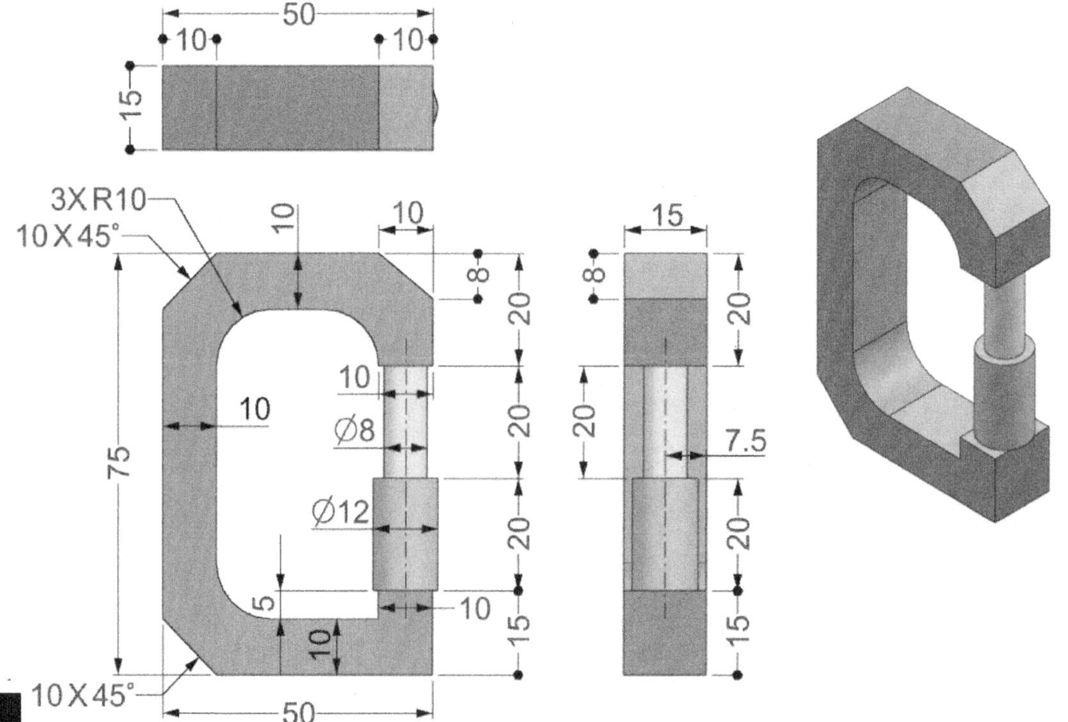
EX-131 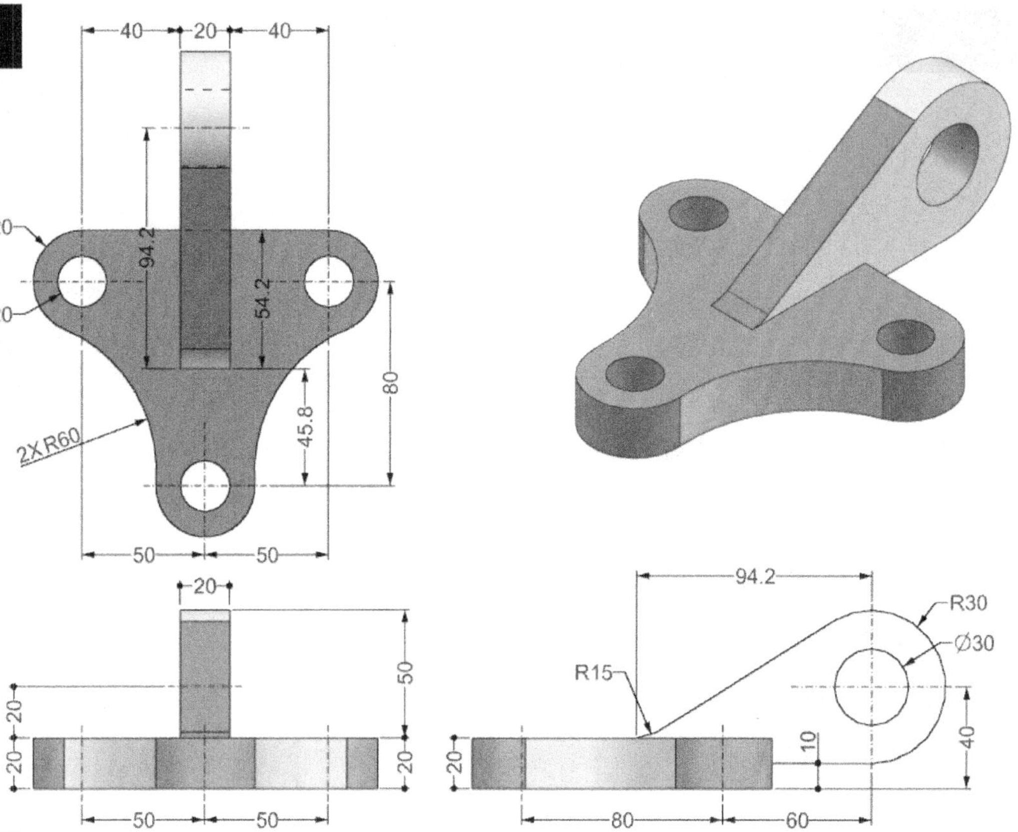
40
20
40
3X R20
94.2
54.2
3X Ø20
2X R60
45.8
80
50
50
20
50
20 20
20
20
50
50
94.2
R30
R15
Ø30
20
10
40
80
60
EX-132
50
10
10
15
3X R10
10 X 45°
10
10
10
8
8
20
20
Ø8
20
20
10
75
Ø12
20
20
10
5
15
10
15
10 X 45°
50
15
7.5

EX-133

Ø30
Ø60
Ø50
50
50
Ø15
Ø10
10
15
Ø15
Ø30
Ø20
36
40
20
18
30
Ø60
Ø15
Ø60
30
18
36
Ø30
Ø30
40
20
10
40
50

EX-134

Ø40
Ø120
Ø60
A
A
80
70
50
R30
2X45°
Ø120
Ø60
Ø93.2
Ø60
Ø120
5
10
10
5
Ø120
Ø60
Ø40
5
2X45°
10
80
Ø93.2
50
70
Ø120
SECTION A-A

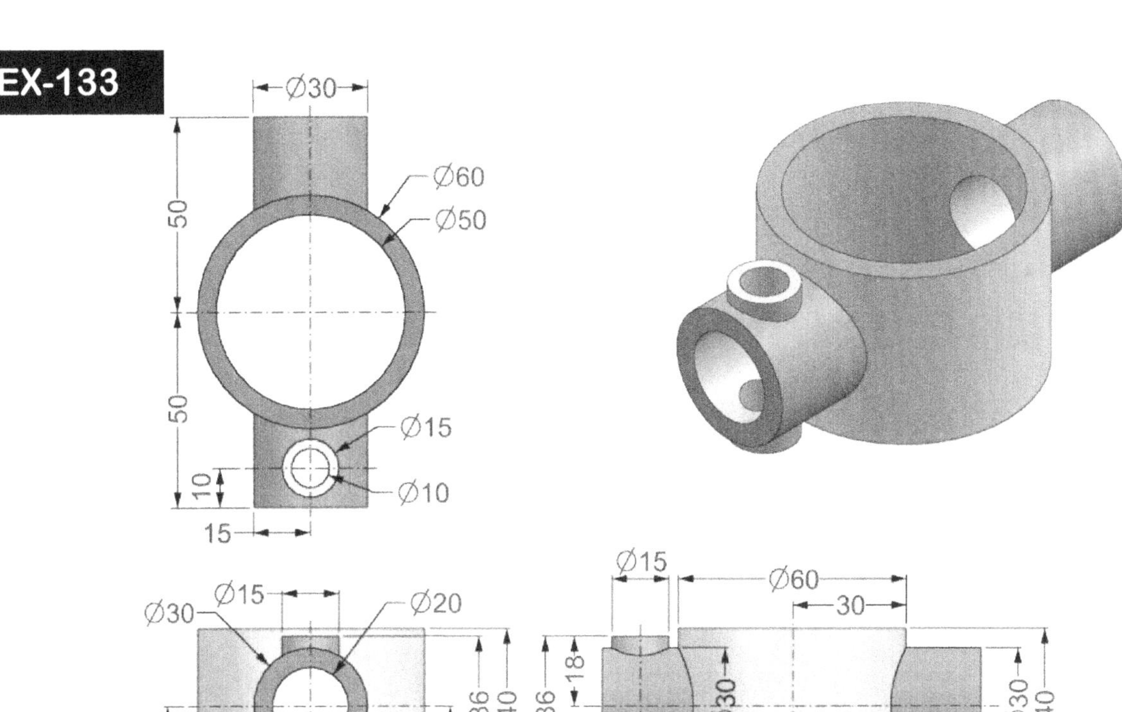

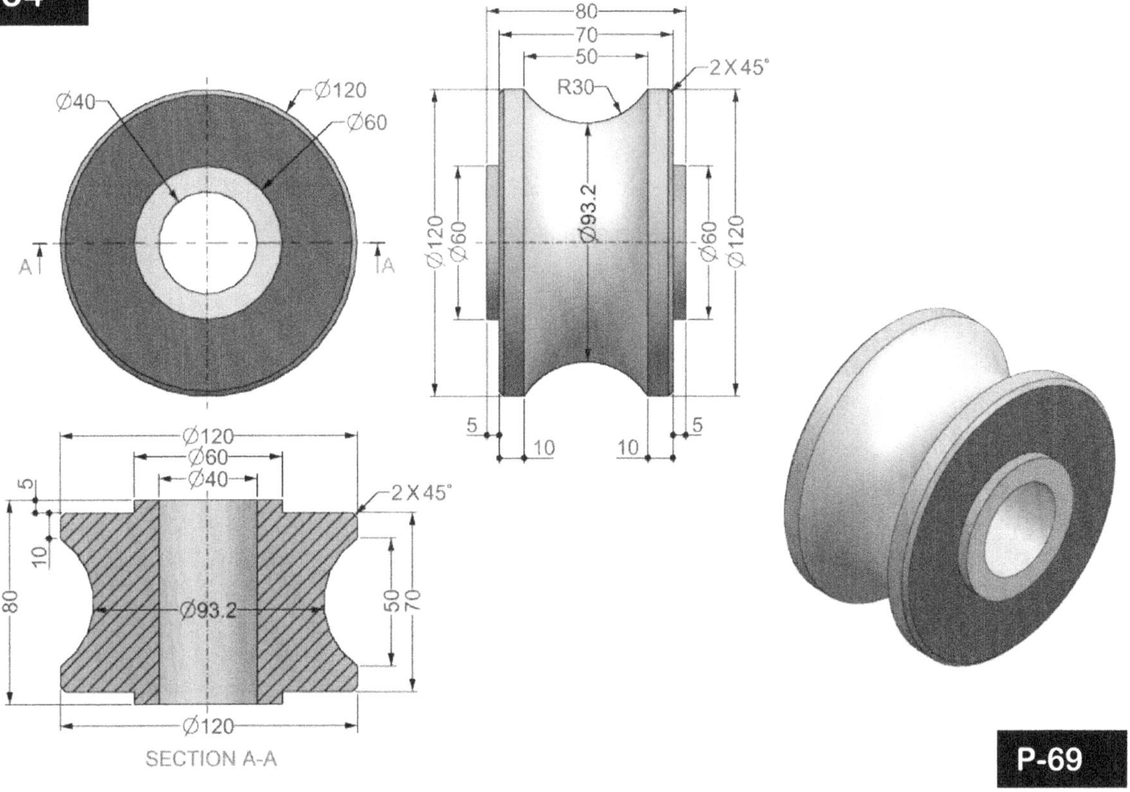

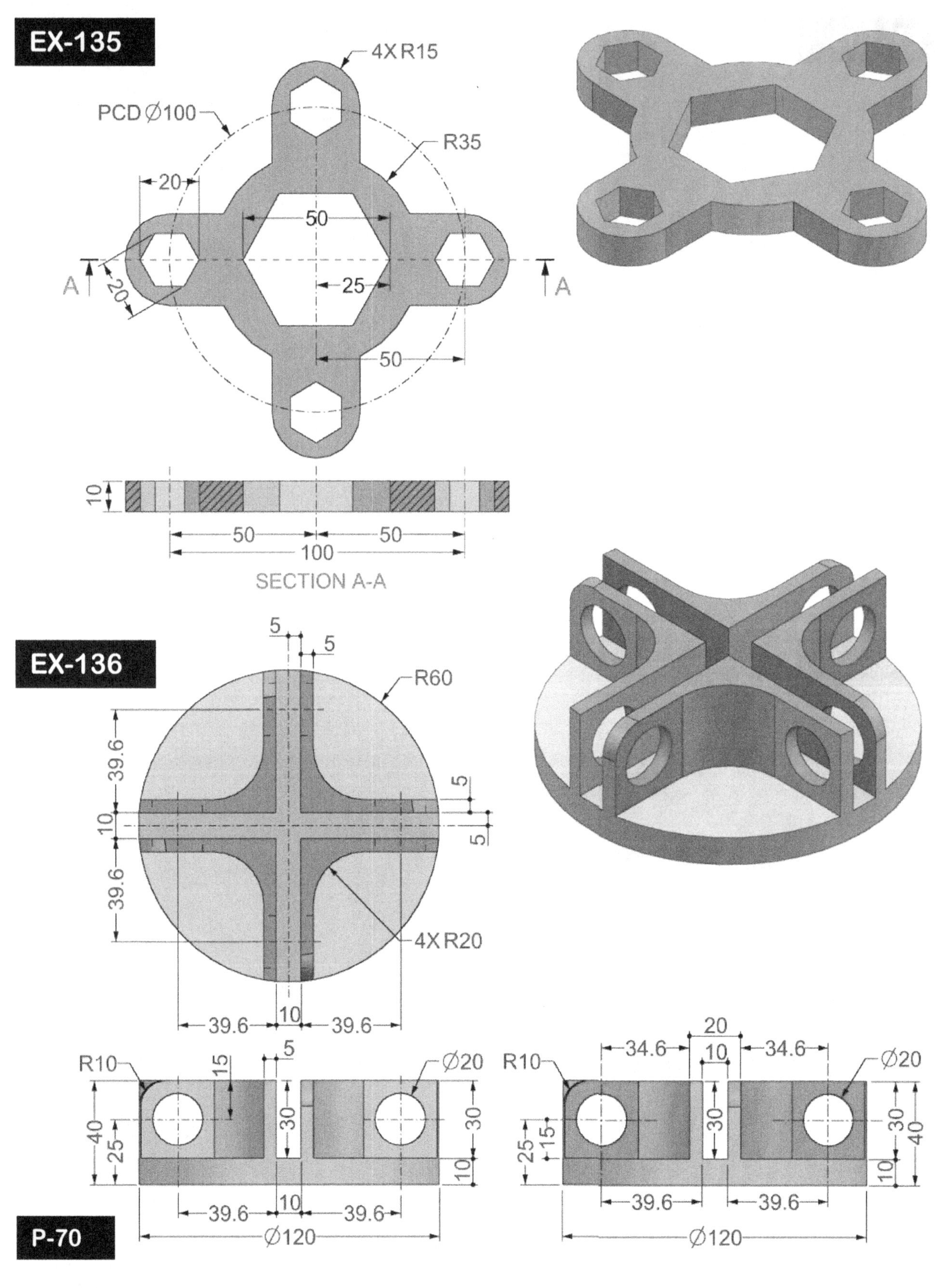

EX-135
4X R15
PCD ⌀100
R35
20
50
25
50
A
A
20
10
50
50
100
SECTION A-A
EX-136
5
5
R60
39.6
39.6
10
5
5
4X R20
39.6
10
39.6
5
R10
15
⌀20
40
25
30
30
10
39.6
10
39.6
⌀120
20
34.6
10
34.6
⌀20
R10
25
15
30
30
40
10
39.6
39.6
⌀120
P-70

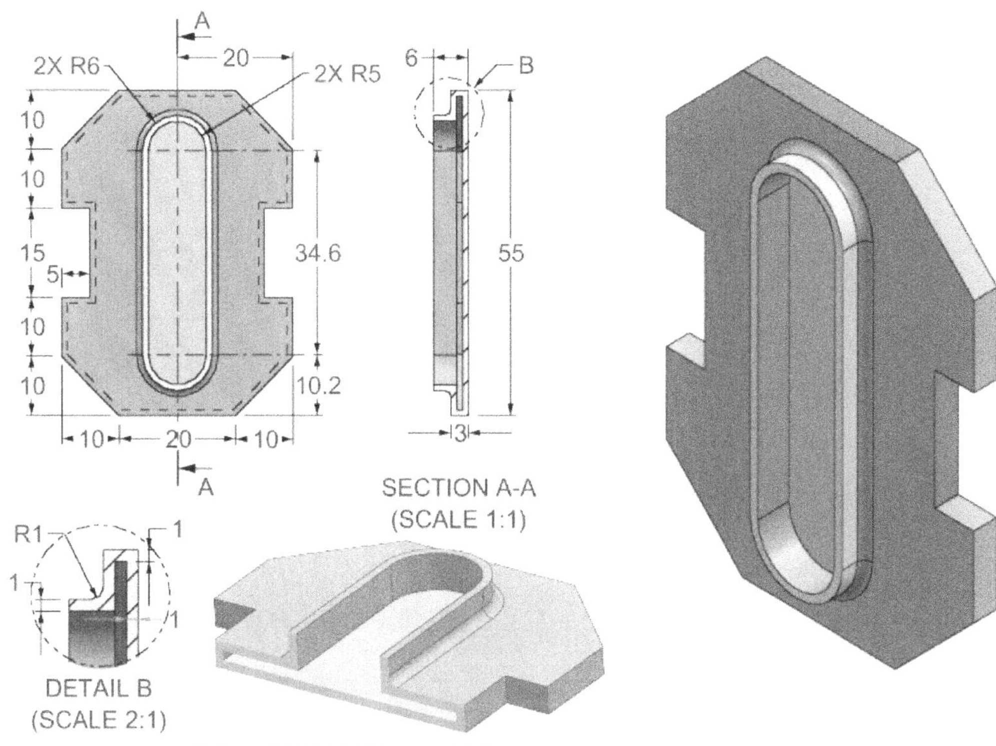

EX-137

2X R6
2X R5
20
10
10
15
5
10
10
34.6
10.2
10
20
10
A
A
6
B
55
3
SECTION A-A
(SCALE 1:1)

R1
1
1
1
DETAIL B
(SCALE 2:1)

SHELL THICKNESS = 1MM
ALL INSIDE WALL THICKNESS

EX-138

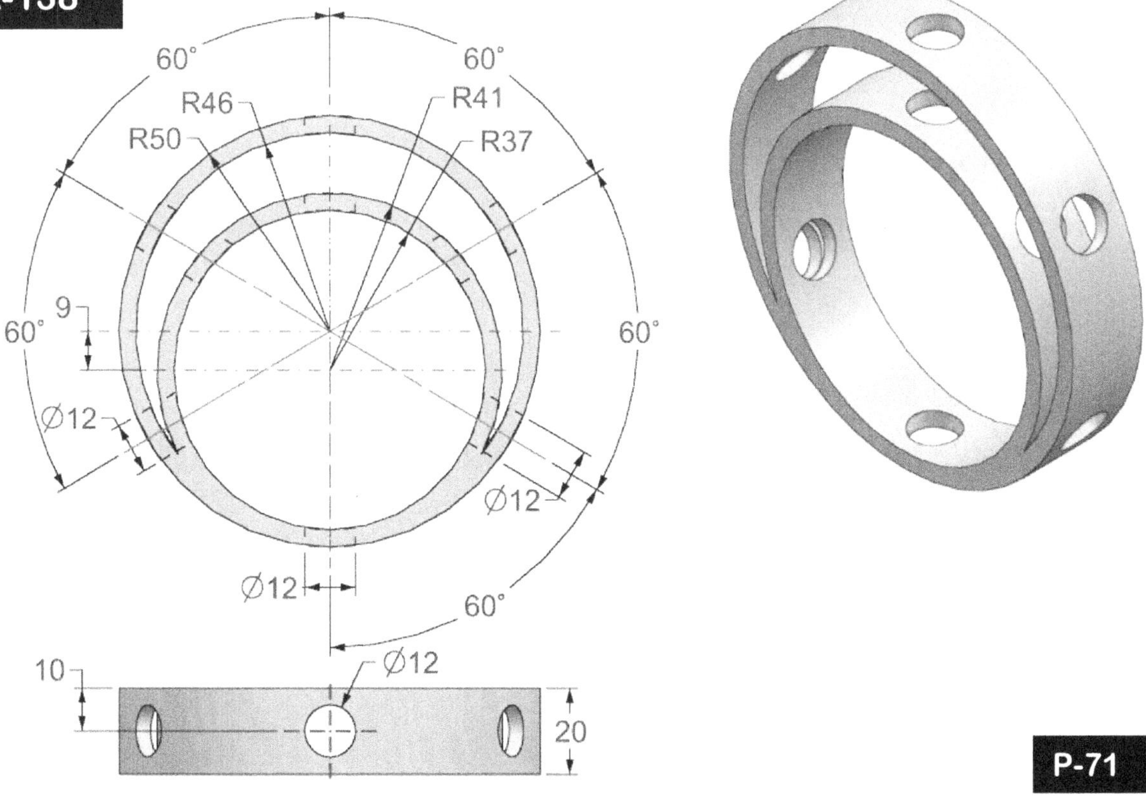

60°
60°
R46
R41
R50
R37
60°
9
60°
Ø12
Ø12
Ø12
60°
Ø12
10
20

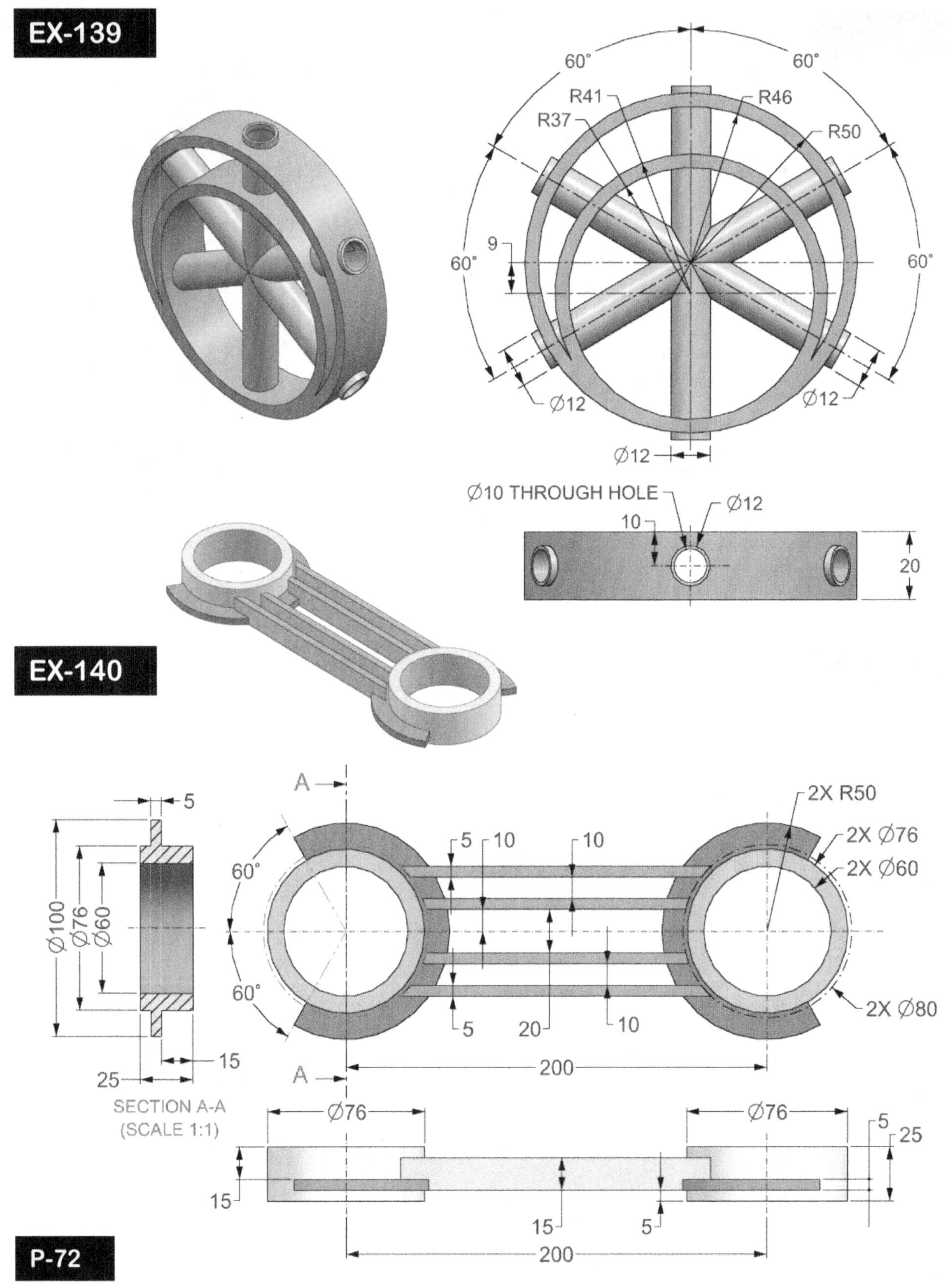

EX-139
60°
60°
R41
R46
R37
R50
60°
60°
9
60°
60°
Ø12
Ø12
Ø12
Ø10 THROUGH HOLE
Ø12
10
20
EX-140
A
5
2X R50
2X Ø76
2X Ø60
60°
5
10
10
5
Ø100
Ø76
Ø60
60°
20
10
2X Ø80
15
5
200
25
A
SECTION A-A
(SCALE 1:1)
Ø76
Ø76
5
25
15
15
5
200
P-72

EX-141 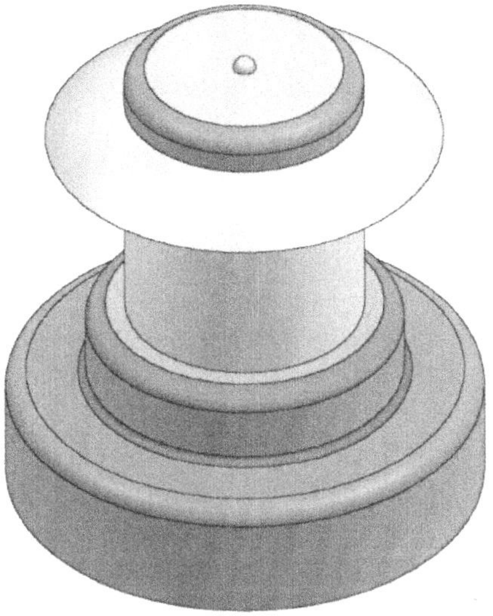 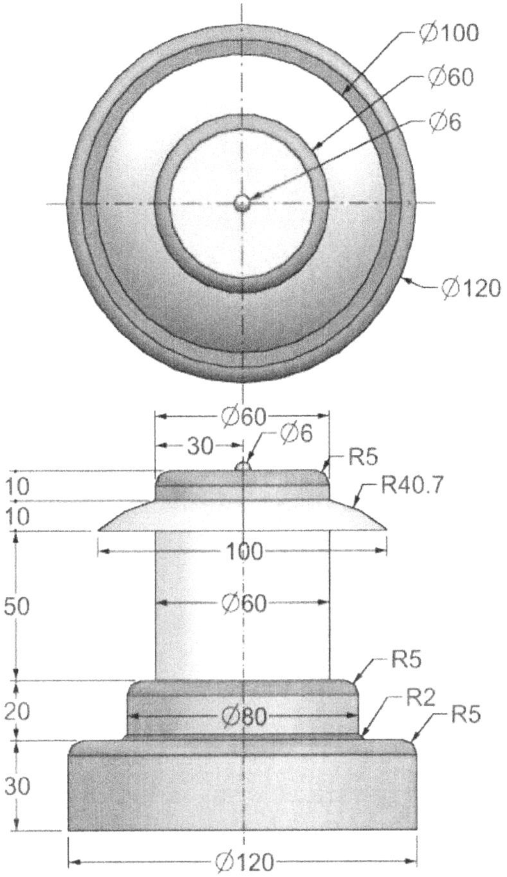
Ø100
Ø60
Ø6
Ø120
Ø60
30
Ø6
R5
10
R40.7
10
100
50
Ø60
R5
20
Ø80
R2
R5
30
R5
Ø120
EX-142 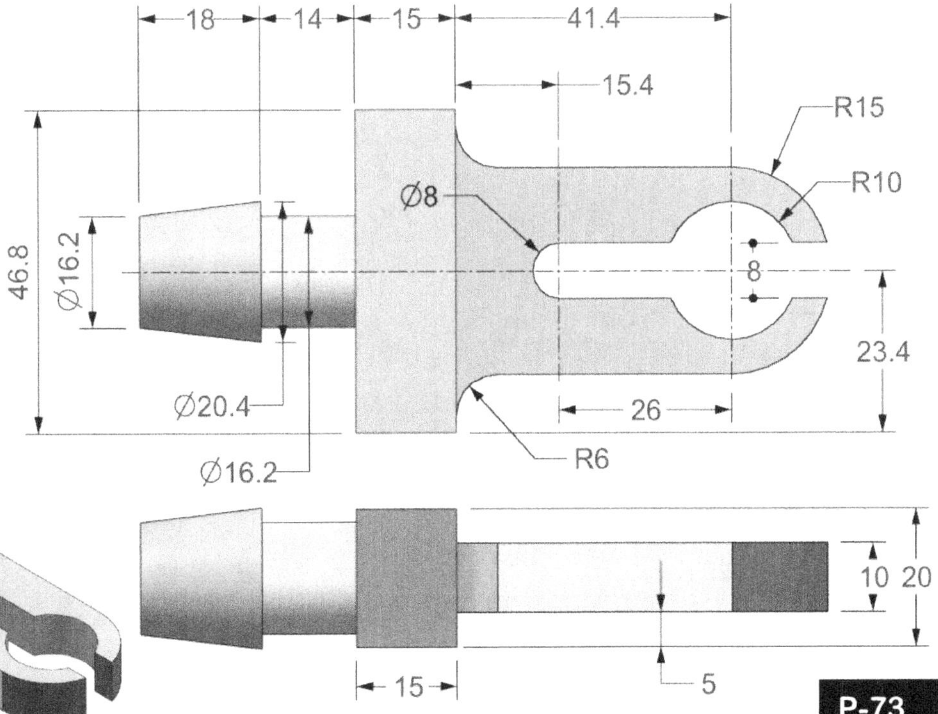
18
14
15
41.4
15.4
R15
Ø8
R10
46.8
Ø16.2
8
23.4
Ø20.4
26
Ø16.2
R6
10 20
15
5
P-73

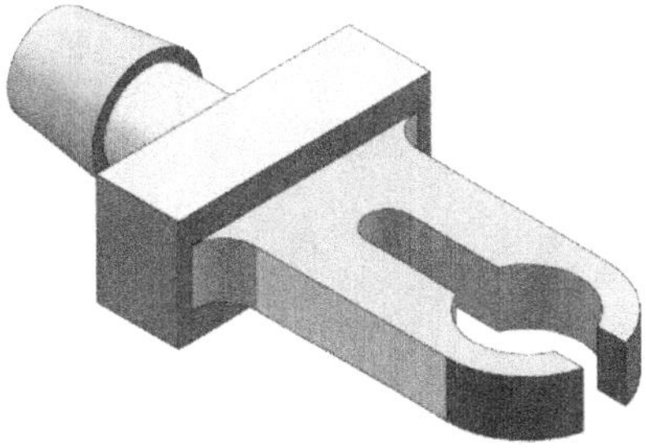

EX-143

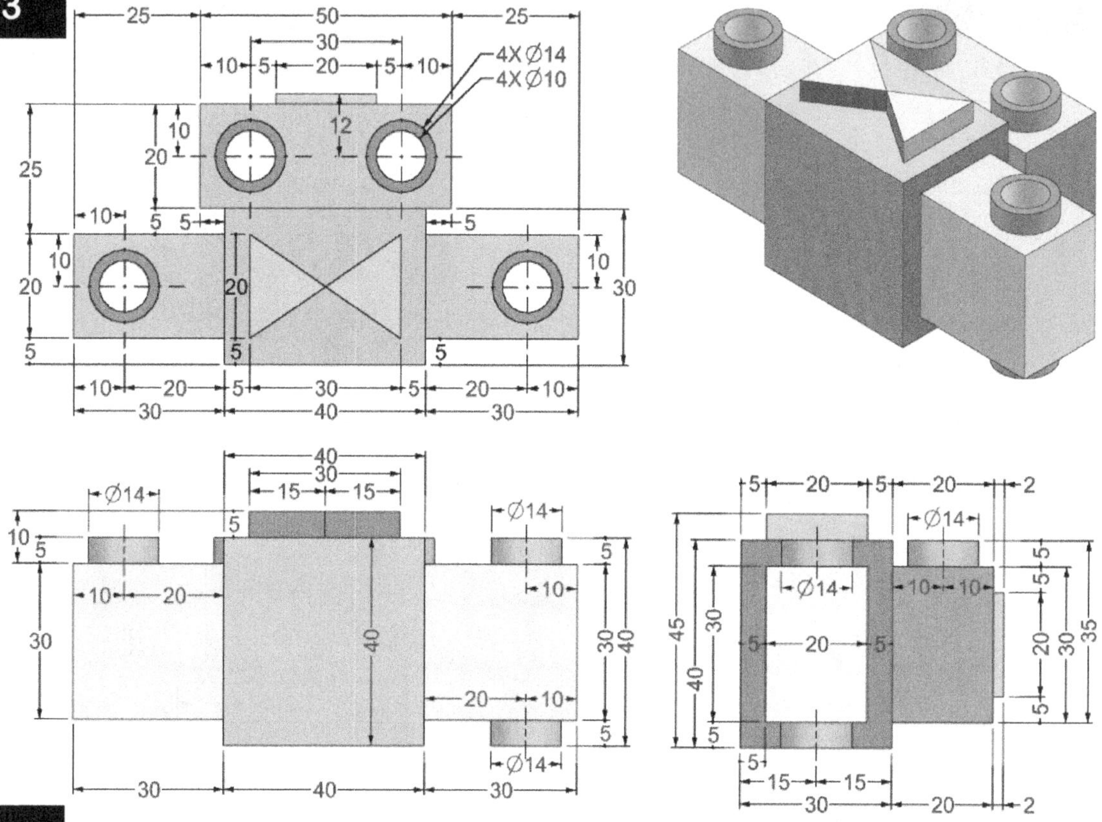

25
50
25
30
10
5
20
5
10
4X Ø14
4X Ø10
10
12
20
25
10
5
5
10
20
10
5
10
30
20
5
5
20
5
30
10
20
5
30
5
20
10
30
40
30

40
30
15
15
Ø14
Ø14
10
5
5
5
10
20
10
30
40
40
30
20
10
30
40
30
Ø14

5
20
5
20
2
Ø14
Ø14
45
40
30
5
20
5
10
10
5
5
5
5
5
20
30
35
15
15
30
20
2

EX-144

20
Ø26
5
Ø60
Ø80
10
5
10
5
SECTION A-A

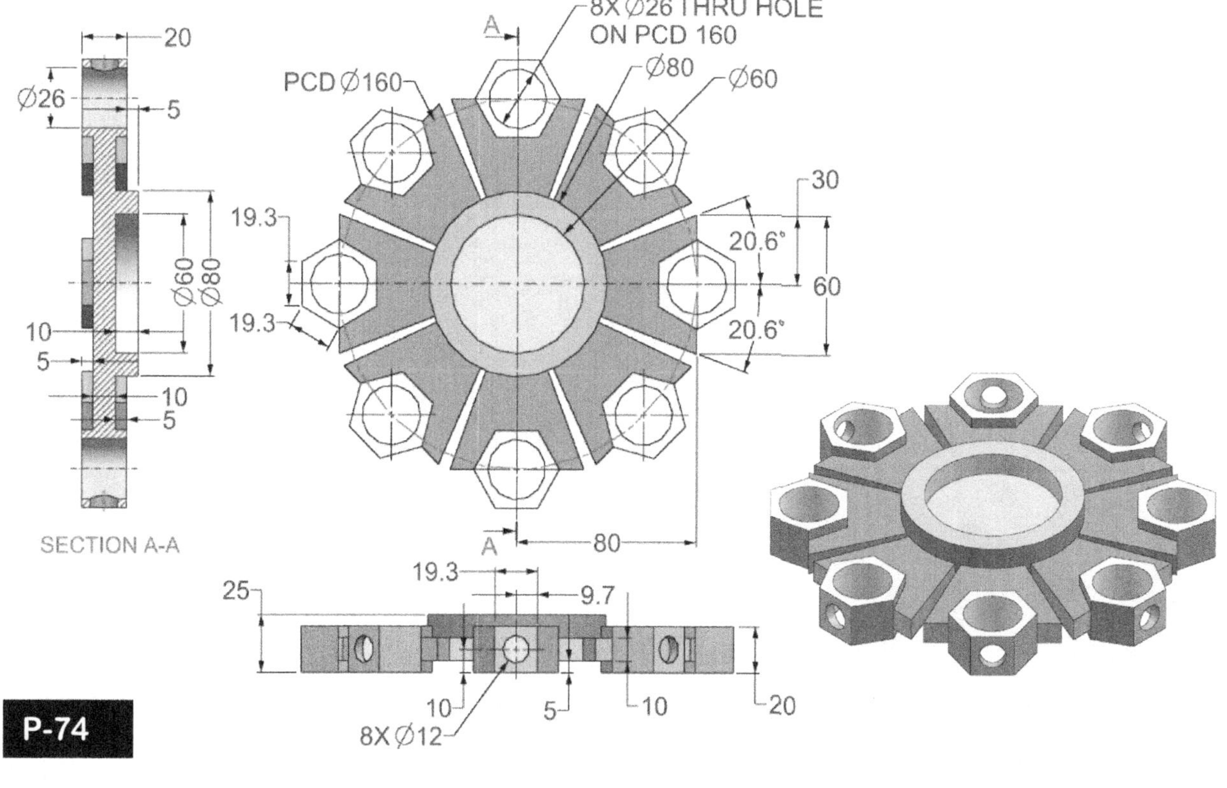

8X Ø26 THRU HOLE
ON PCD 160
PCD Ø160
Ø80
Ø60
A
A
19.3
19.3
30
20.6°
60
20.6°
80

19.3
9.7
25
10
5
10
20
8X Ø12

P-74

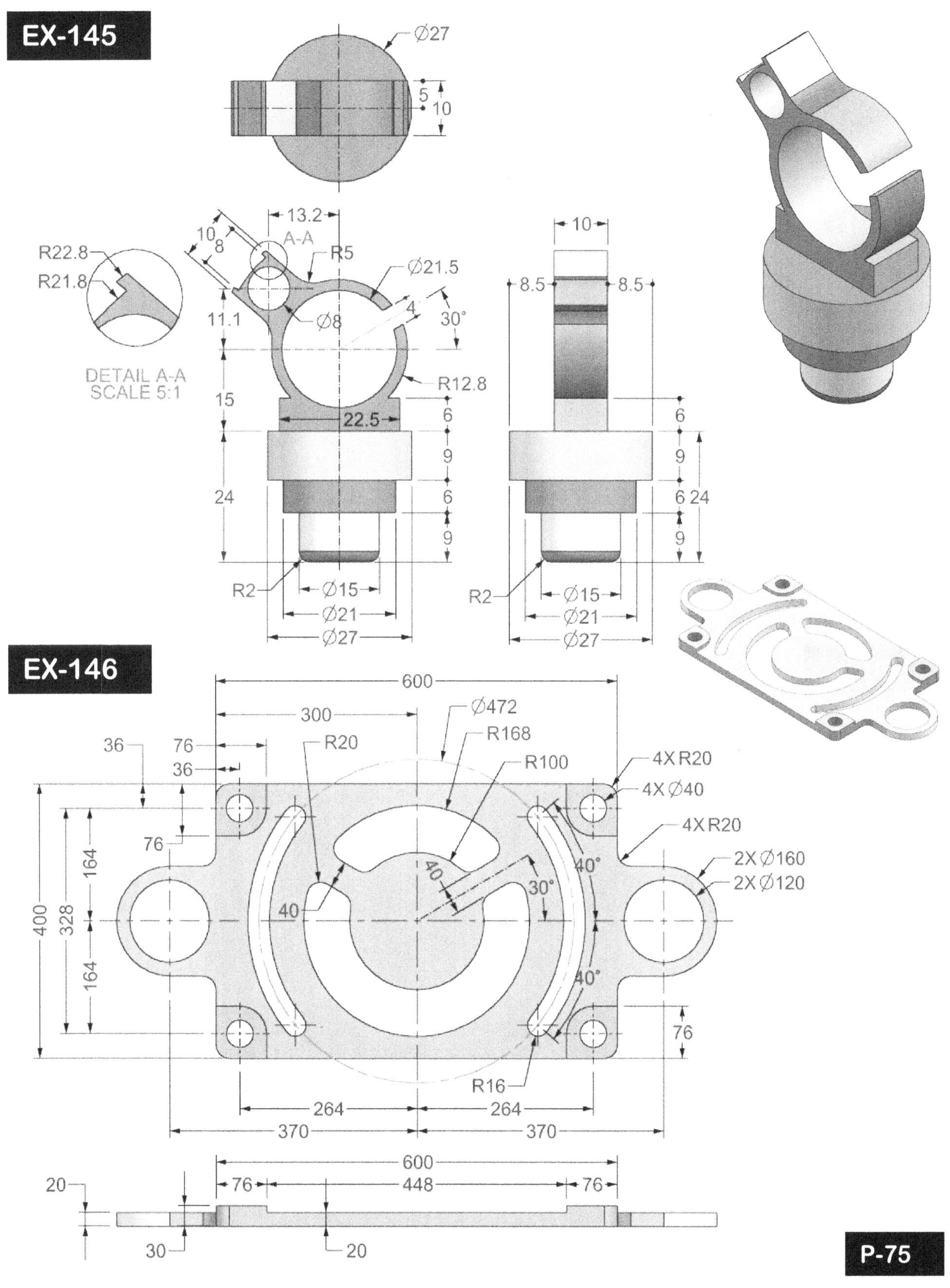

EX-145

Ø27
5
10

R22.8
R21.8

DETAIL A-A
SCALE 5:1

13.2
10
8
A-A
R5
Ø21.5
11.1
Ø8
4
30°
15
R12.8
22.5
6
9
6
9
24
R2
Ø15
Ø21
Ø27

10
8.5
8.5
6
9
6
9
6 24
R2
Ø15
Ø21
Ø27

EX-146

600
300
Ø472
R168
R100
4X R20
4X Ø40
4X R20
2X Ø160
2X Ø120
36
76
R20
36
76
164
328
400
164
40
40
30°
40°
40°
40°
R16
264
264
370
370

600
76
448
76
20
30
20

EX-147

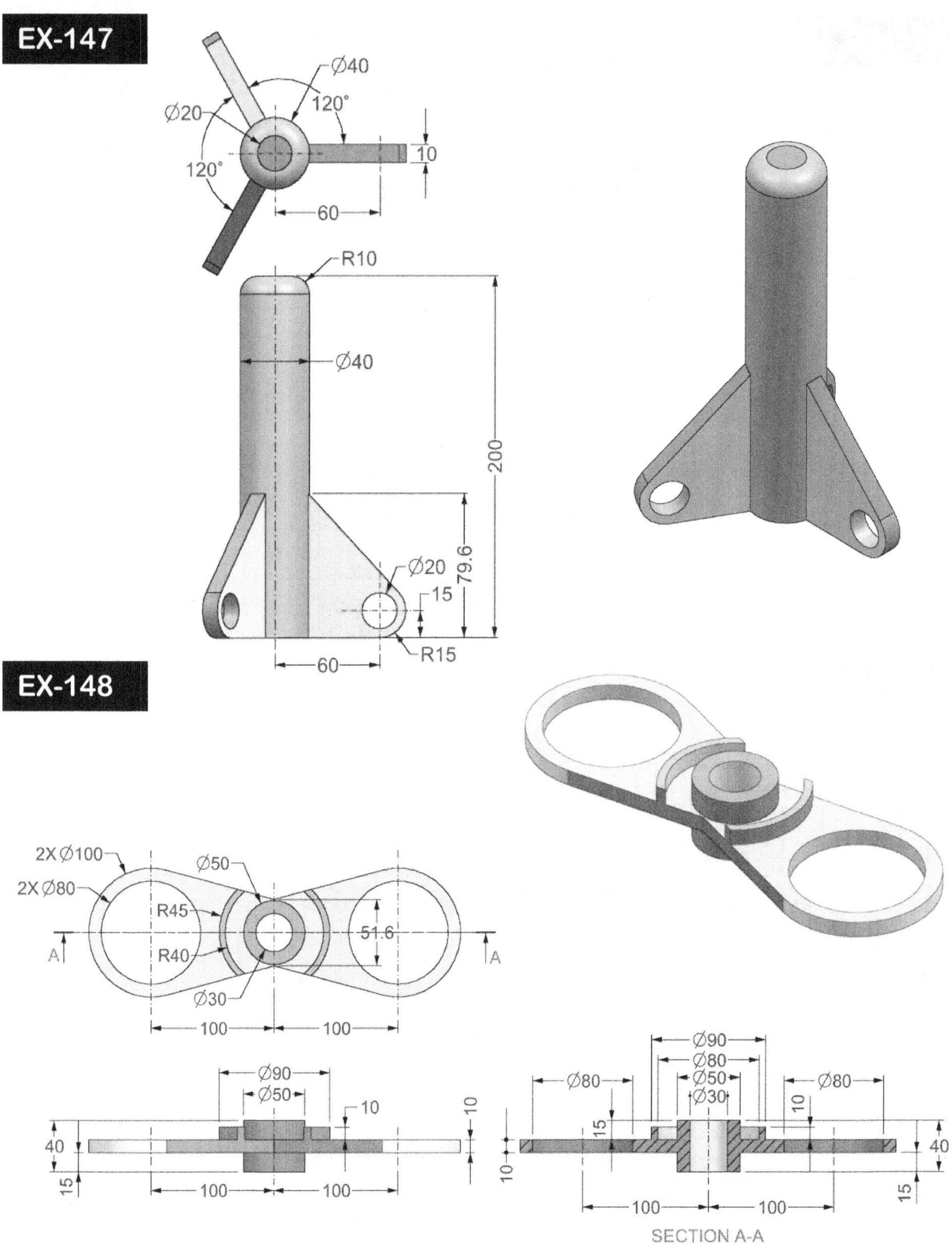

EX-147
Ø40
Ø20
120°
120°
10
60
R10
Ø40
200
79.6
Ø20
15
60
R15
EX-148
2X Ø100
2X Ø80
Ø50
R45
R40
Ø30
51.6
100
100
Ø90
Ø50
10
10
40
15
100
100
Ø90
Ø80
Ø50
Ø30
Ø80
Ø80
15
10
10
40
15
100
100
SECTION A-A
A
A
P-76

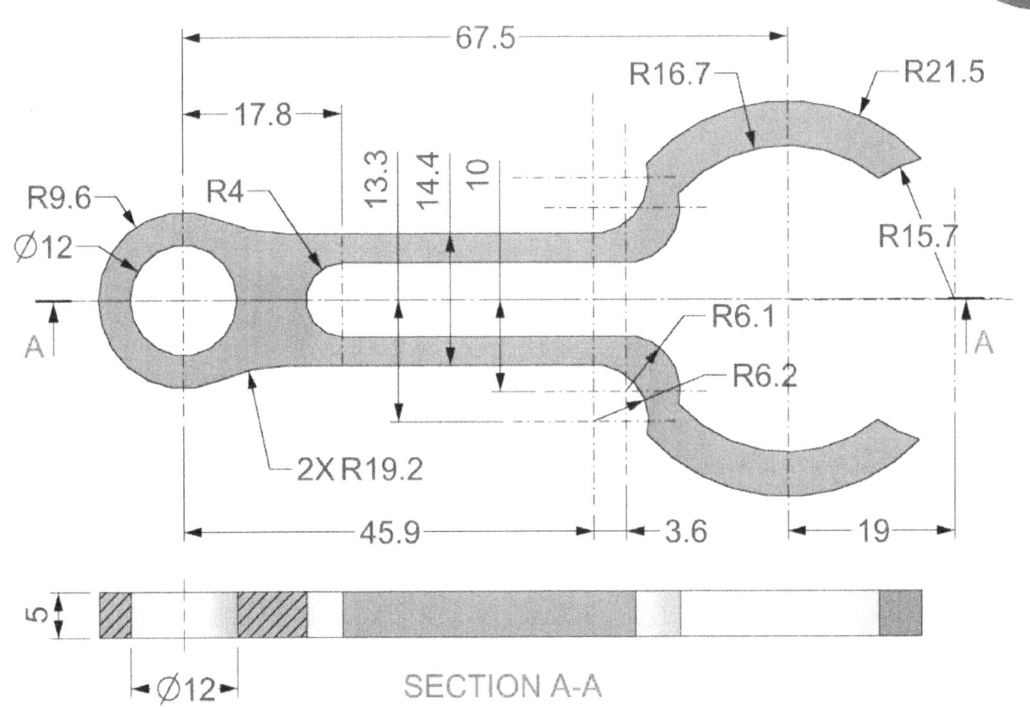

EX-149 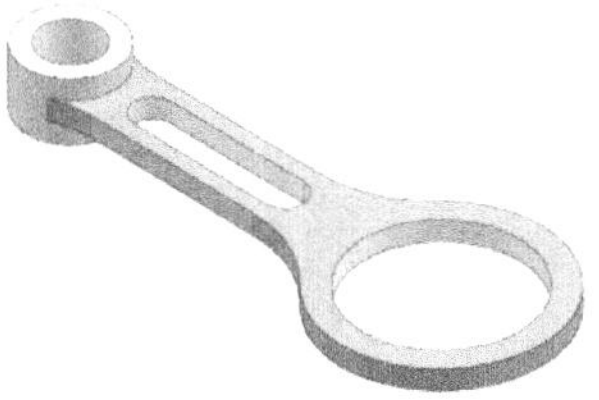 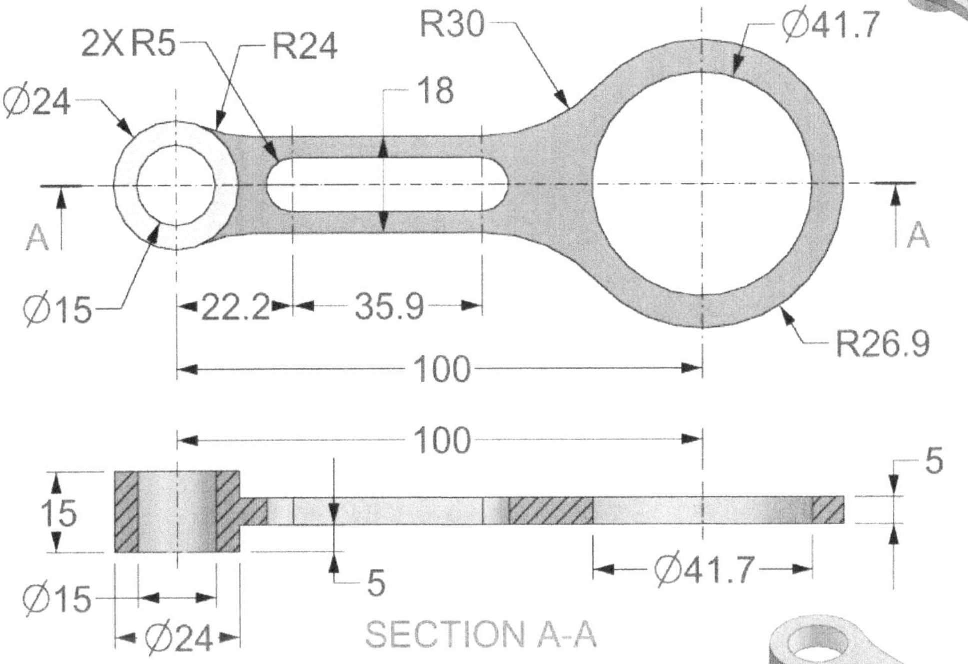
2X R5
R24
R30
Ø41.7
18
Ø24
R26.9
Ø15
22.2
35.9
100
100
15
5
Ø15
5
Ø24
Ø41.7
SECTION A-A
A
A
EX-150 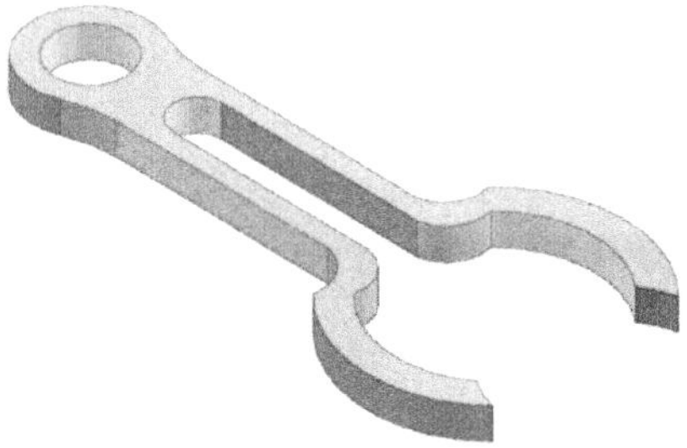
67.5
R16.7
R21.5
17.8
13.3
14.4
10
R9.6
R4
R15.7
Ø12
R6.1
R6.2
2X R19.2
45.9
3.6
19
A
A
5
Ø12
SECTION A-A
P-77

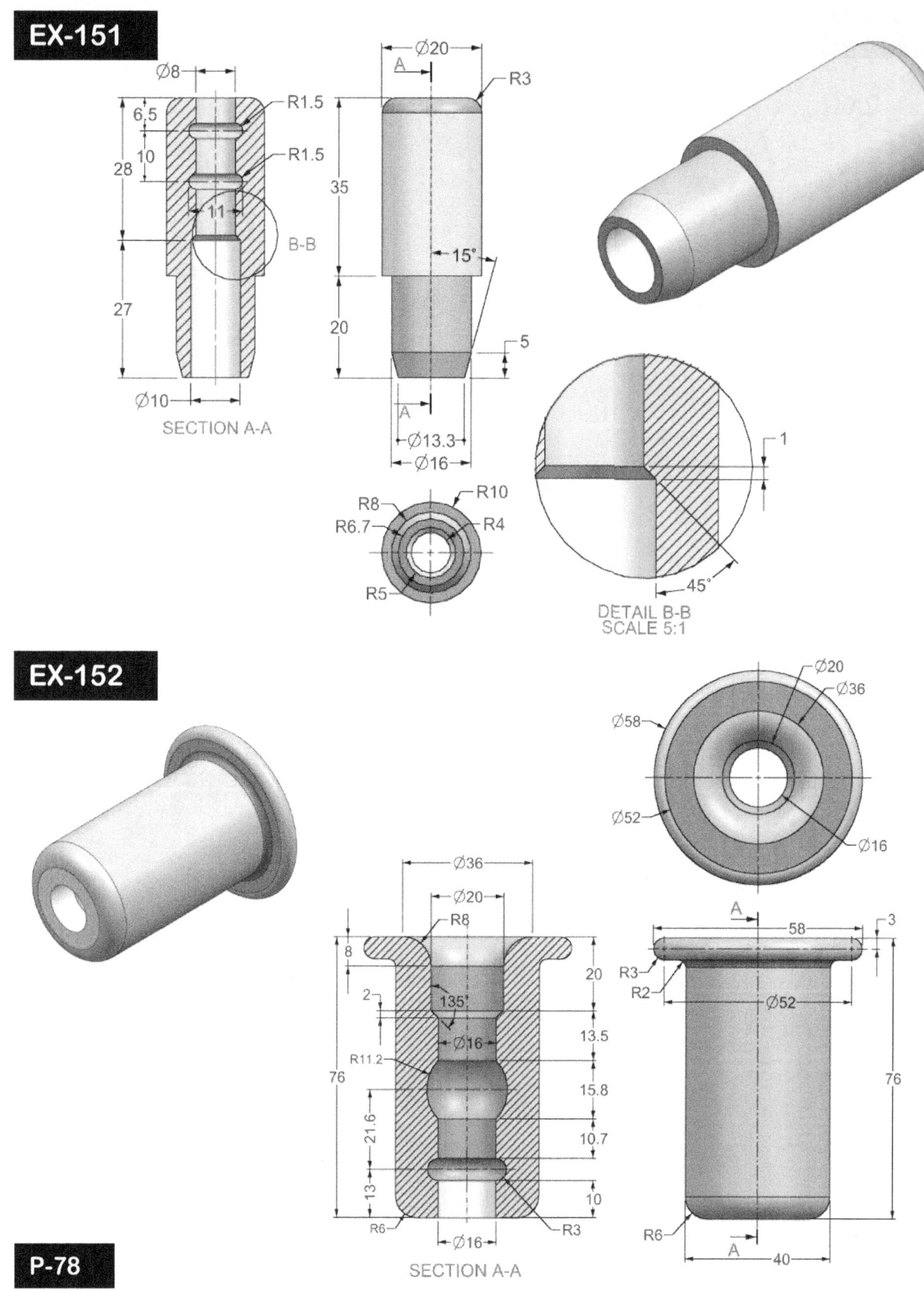

EX-151
Ø8
6.5
10
28
11
27
R1.5
R1.5
B-B
Ø10
SECTION A-A
Ø20
A
R3
35
15°
20
5
A
Ø13.3
Ø16
R8
R10
R6.7
R4
R5
1
45°
DETAIL B-B
SCALE 5:1
EX-152
Ø20
Ø36
Ø58
Ø52
Ø16
Ø36
Ø20
R8
8
20
2
135°
Ø16
13.5
R11.2
15.8
76
21.6
10.7
13
10
R6
R3
Ø16
SECTION A-A
A
58
3
R3
R2
Ø52
76
R6
A
40
P-78

EX-153

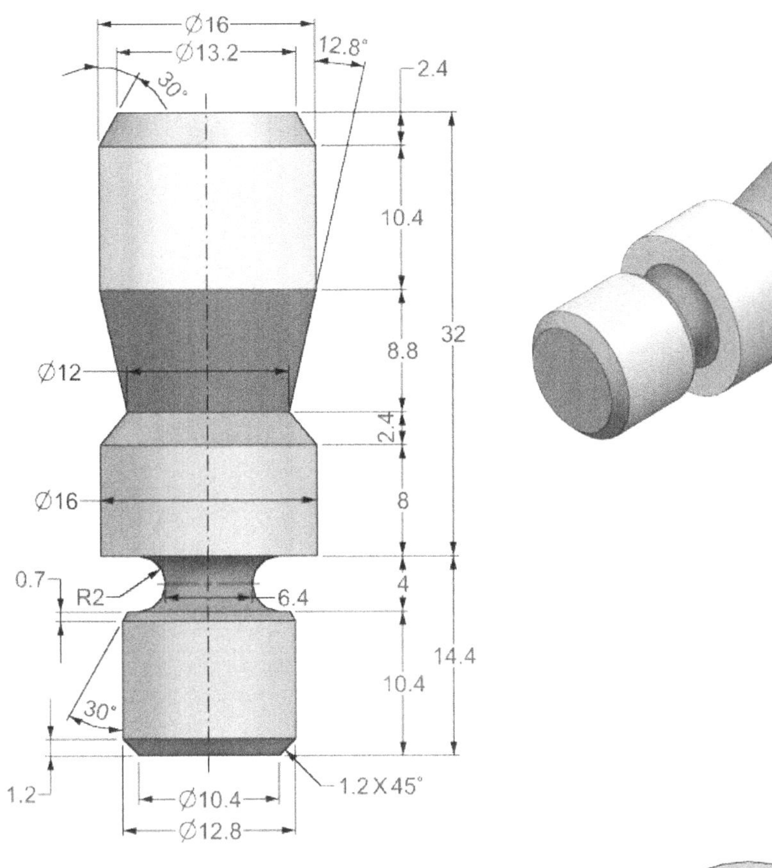

Ø16
Ø13.2
12.8°
30°
2.4
10.4
32
8.8
Ø12
2.4
Ø16
8
0.7
R2
6.4
4
14.4
30°
10.4
1.2
Ø10.4
1.2 X 45°
Ø12.8

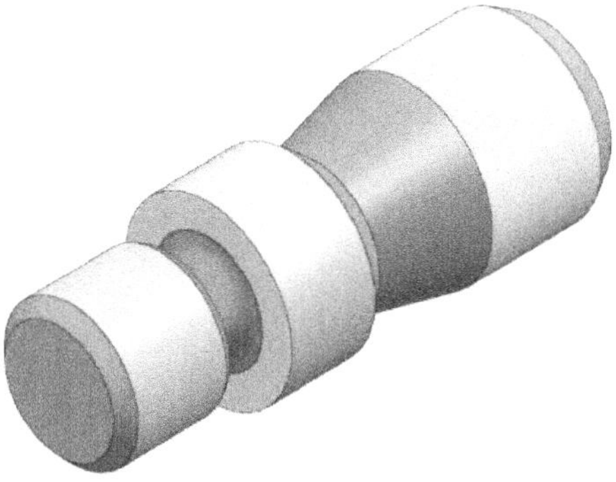

EX-154

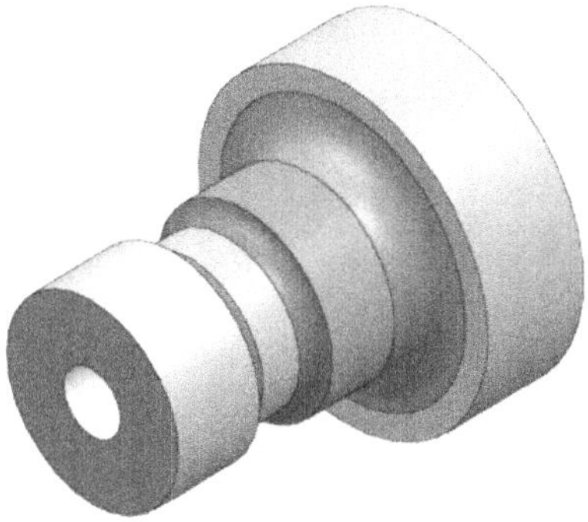

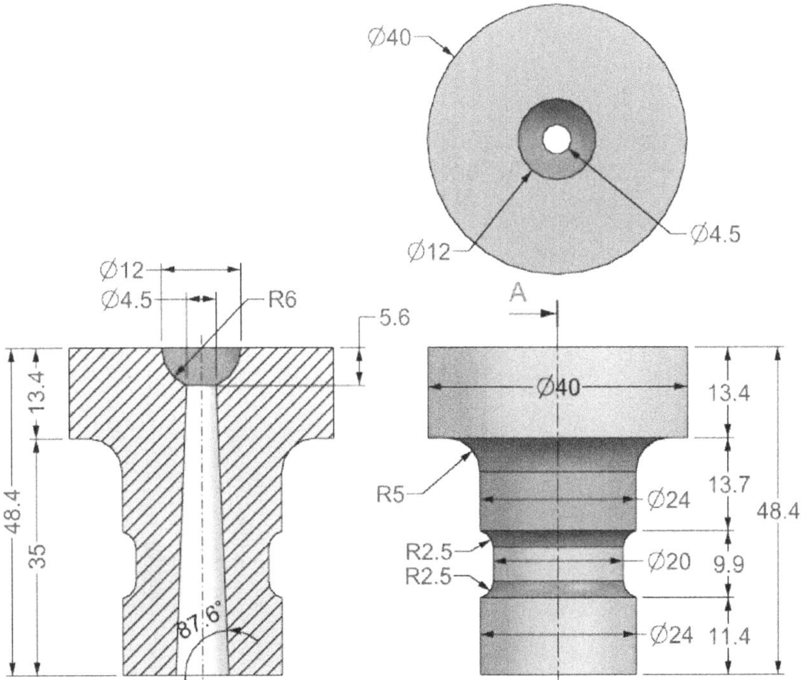

Ø40
Ø12
Ø4.5
A
Ø12
Ø4.5
R6
5.6
Ø40
13.4
13.4
R5
13.7
Ø24
35
48.4
R2.5
Ø20
R2.5
9.9
87.6°
Ø24
11.4
Ø8
SECTION A-A
A
P-79

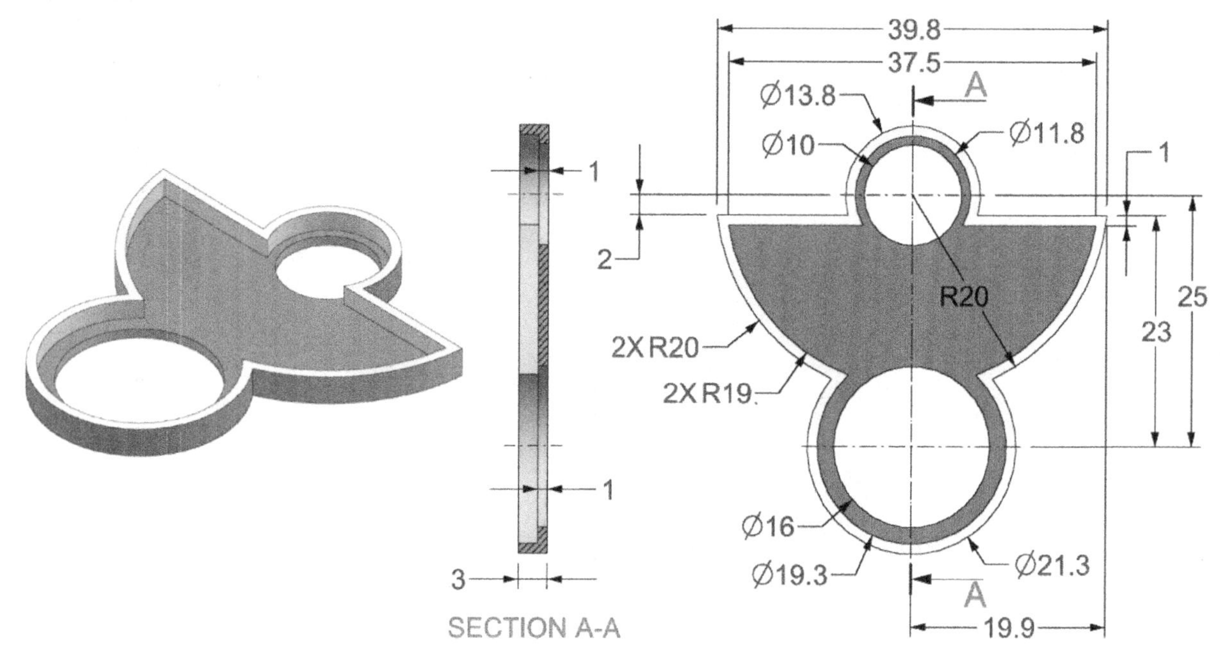

EX-155
39.8
37.5
Ø13.8
Ø10
Ø11.8
A
1
2
R20
25
23
2X R20
2X R19
Ø16
Ø19.3
Ø21.3
A
19.9
1
3
SECTION A-A
EX-156
Ø20 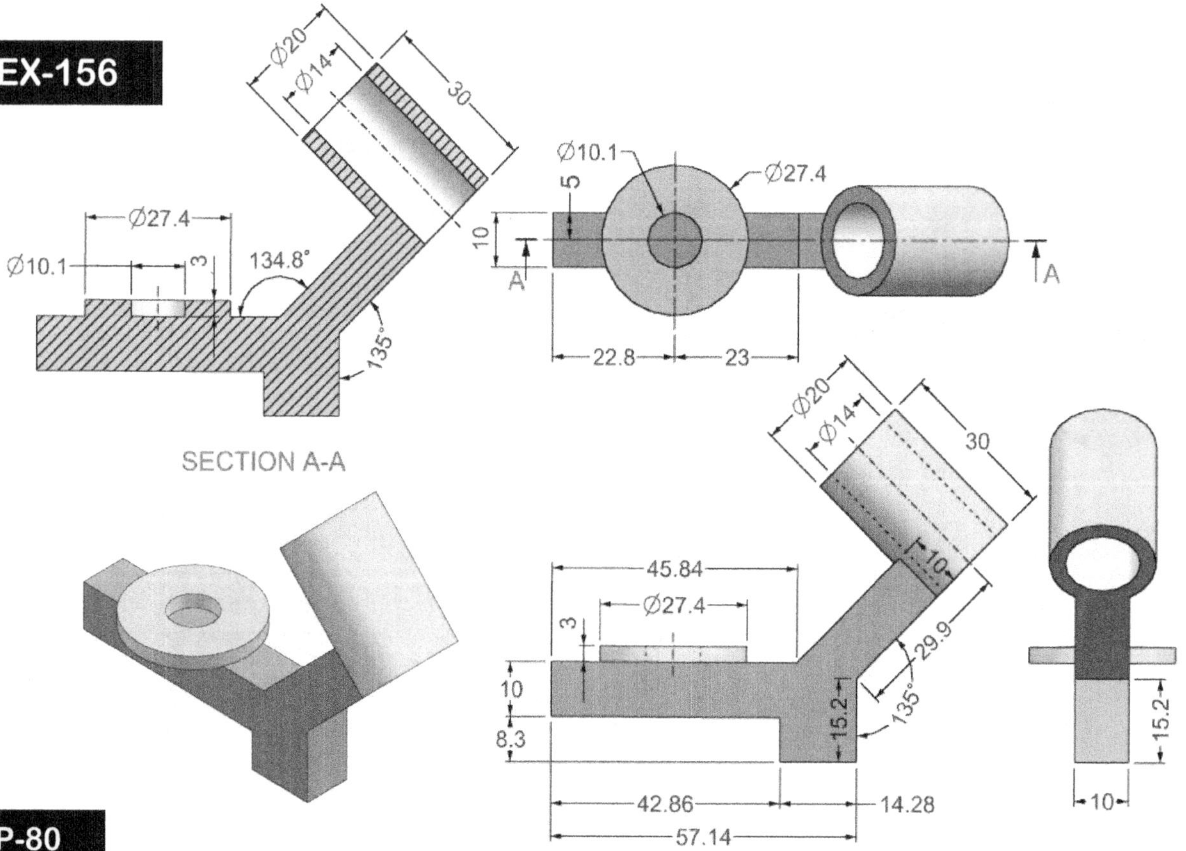
Ø14
30
Ø27.4
Ø10.1
3
134.8°
135°
SECTION A-A
Ø10.1
5
Ø27.4
10
A
22.8
23
A
Ø20
Ø14
30
10
29.9
135°
15.2
45.84
Ø27.4
3
10
8.3
42.86
14.28
57.14
15.2
10
P-80

P-81

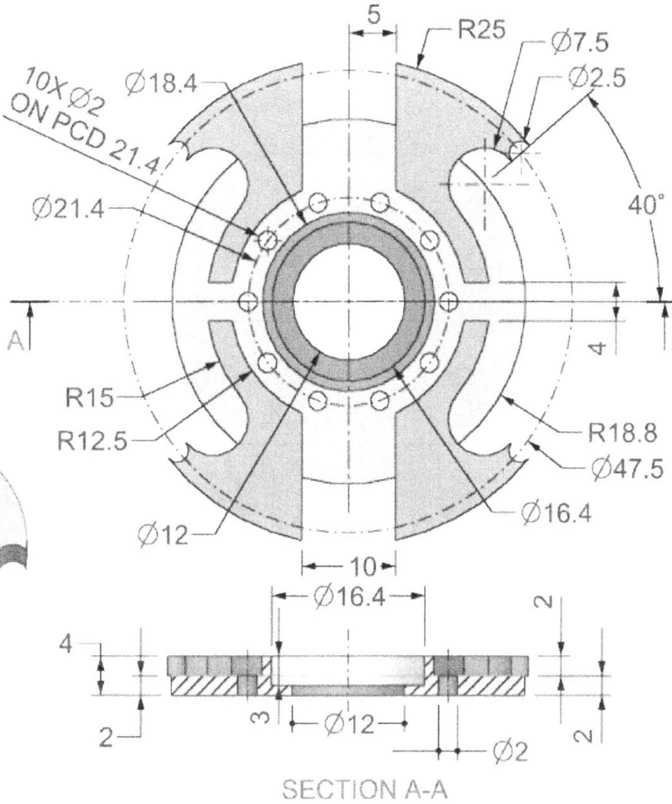

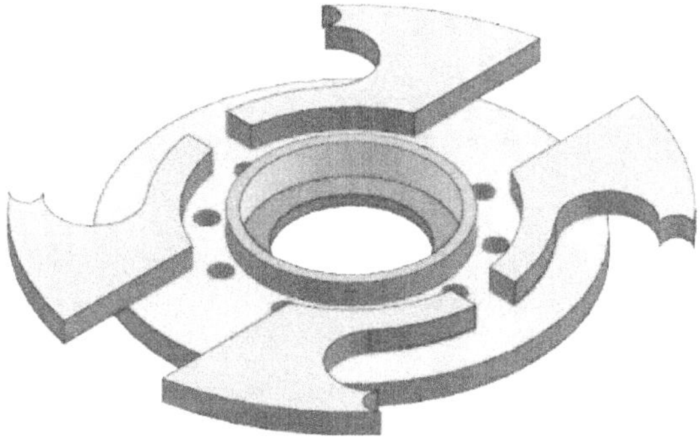

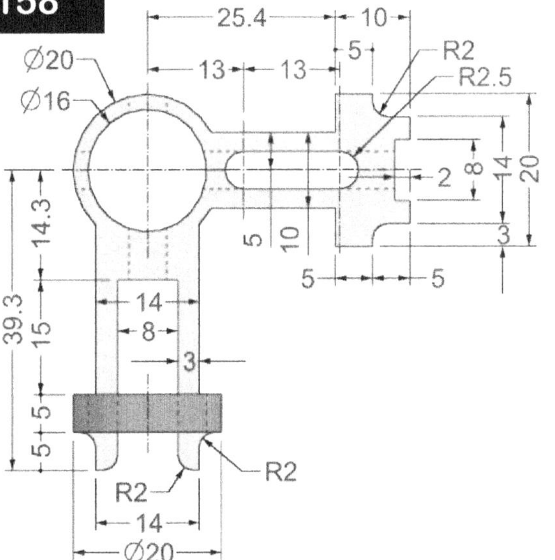

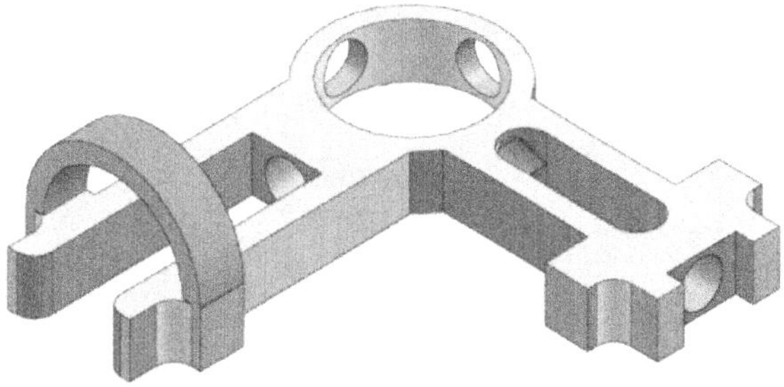

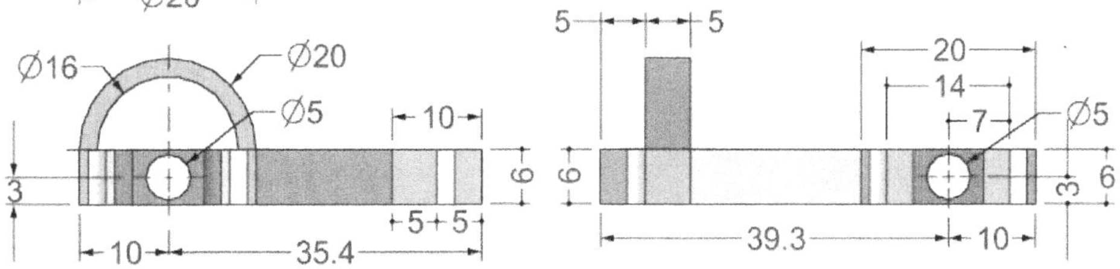

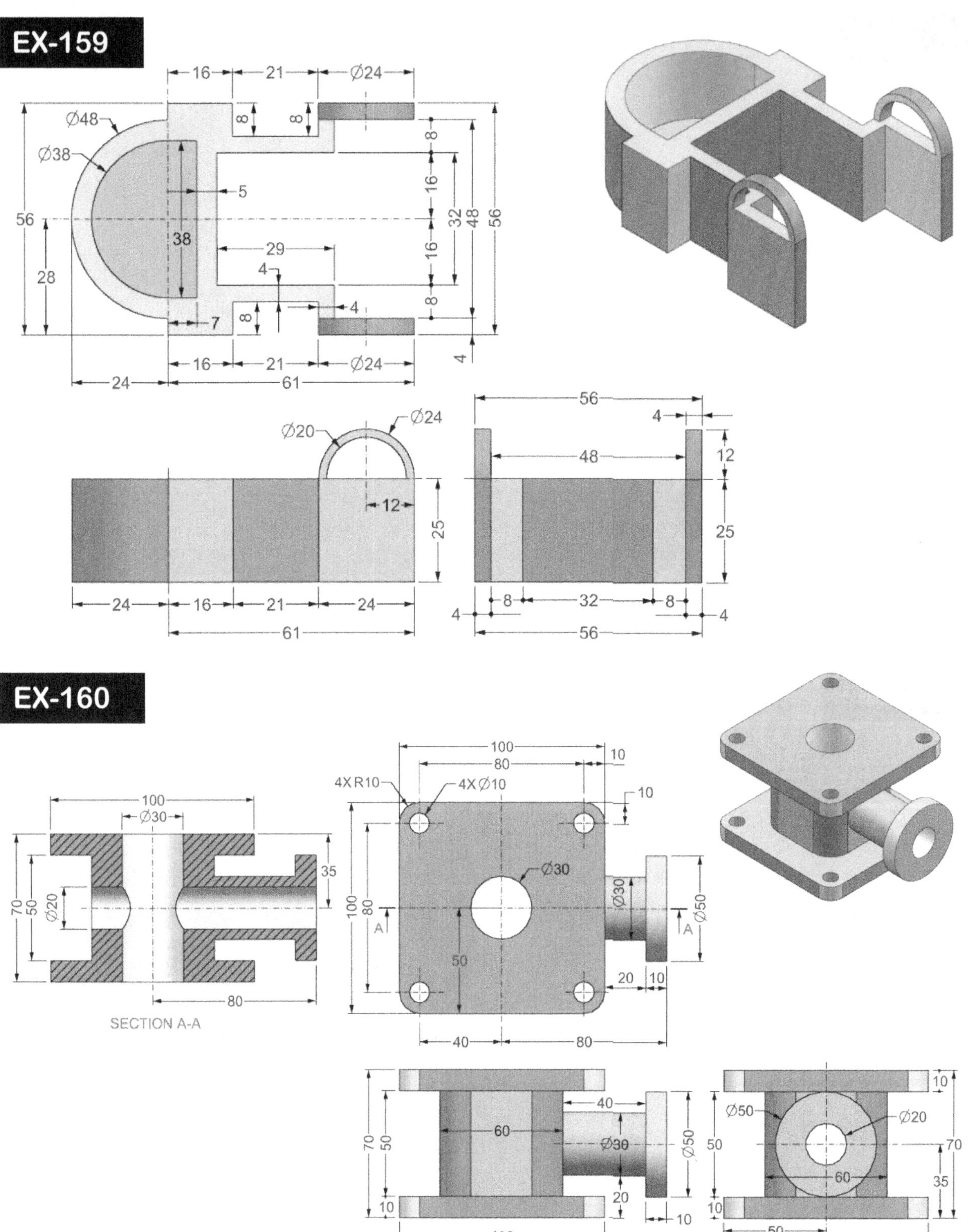

EX-160
SECTION A-A
4X R10
4X Ø10
P-82

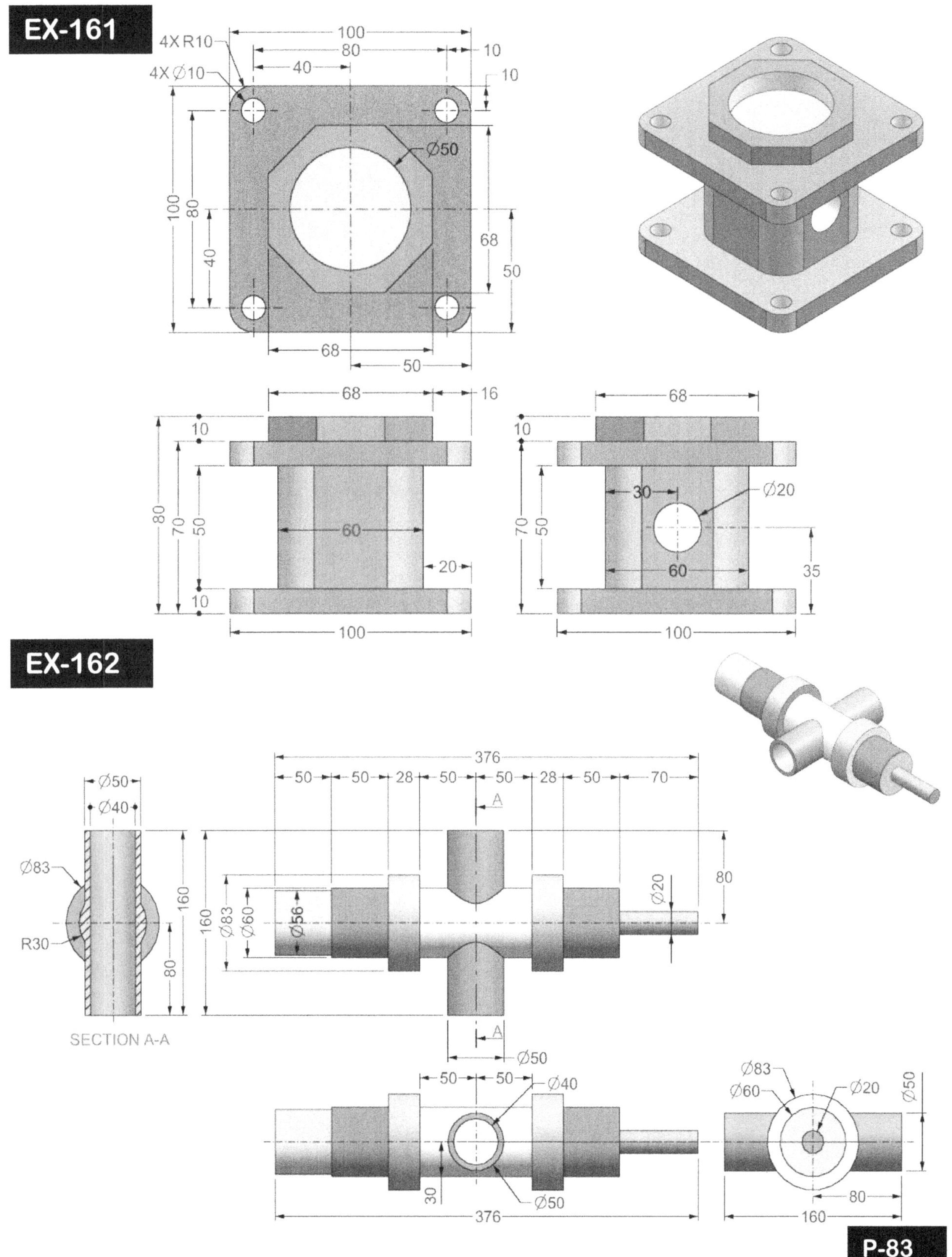

EX-161
4X R10
4X Ø10
100
80
40
10
10
100
80
40
Ø50
68
50
68
50
EX-162
68
16
10
80
70
50
60
20
10
100
68
10
80
70
50
30
Ø20
60
35
100
376
50
50
28
50
50
28
50
70
A
Ø50
Ø40
Ø83
R30
160
160
80
Ø83
Ø60
Ø56
Ø20
80
A
SECTION A-A
Ø50
50
50
Ø40
30
Ø50
376
Ø83
Ø60
Ø20
Ø50
80
160
P-83

EX-163

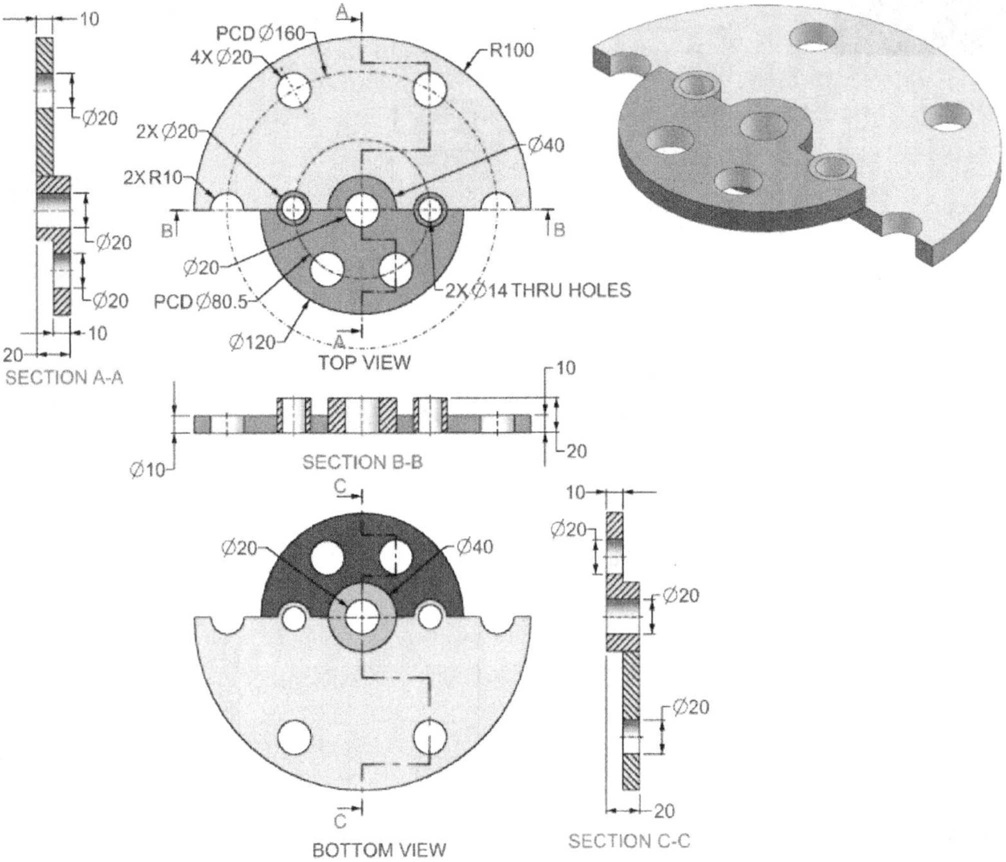

10
Ø20

Ø20

PCD Ø160
4X Ø20
A
R100
2X Ø20
2X R10
Ø40
B
B
Ø20
PCD Ø80.5
2X Ø14 THRU HOLES
Ø120
A
TOP VIEW

Ø20
Ø20
10
20
Ø10
SECTION A-A

10
20
SECTION B-B

C
Ø20
Ø40
Ø20
C
BOTTOM VIEW

10
Ø20
Ø20
Ø20
20
SECTION C-C

EX-164

68
28.2
4X R10
4X Ø10
10
Ø50
Ø30
68
28.2
80
100
A
A
40
10
10
40
40
10
80
100

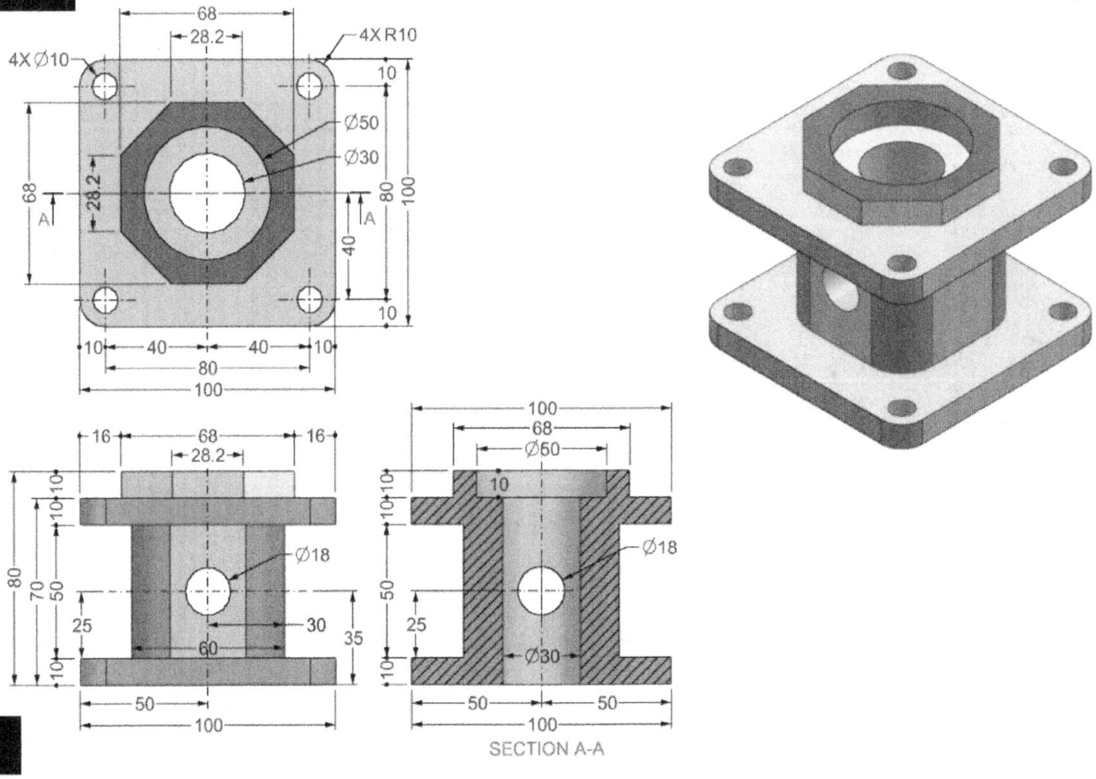

16
68
16
28.2
10 10
Ø18
80
70
50
25
30
60
35
50
100

100
68
Ø50
10 10
10
50
25
Ø18
Ø30
50
50
100
SECTION A-A

P-84

EX-165

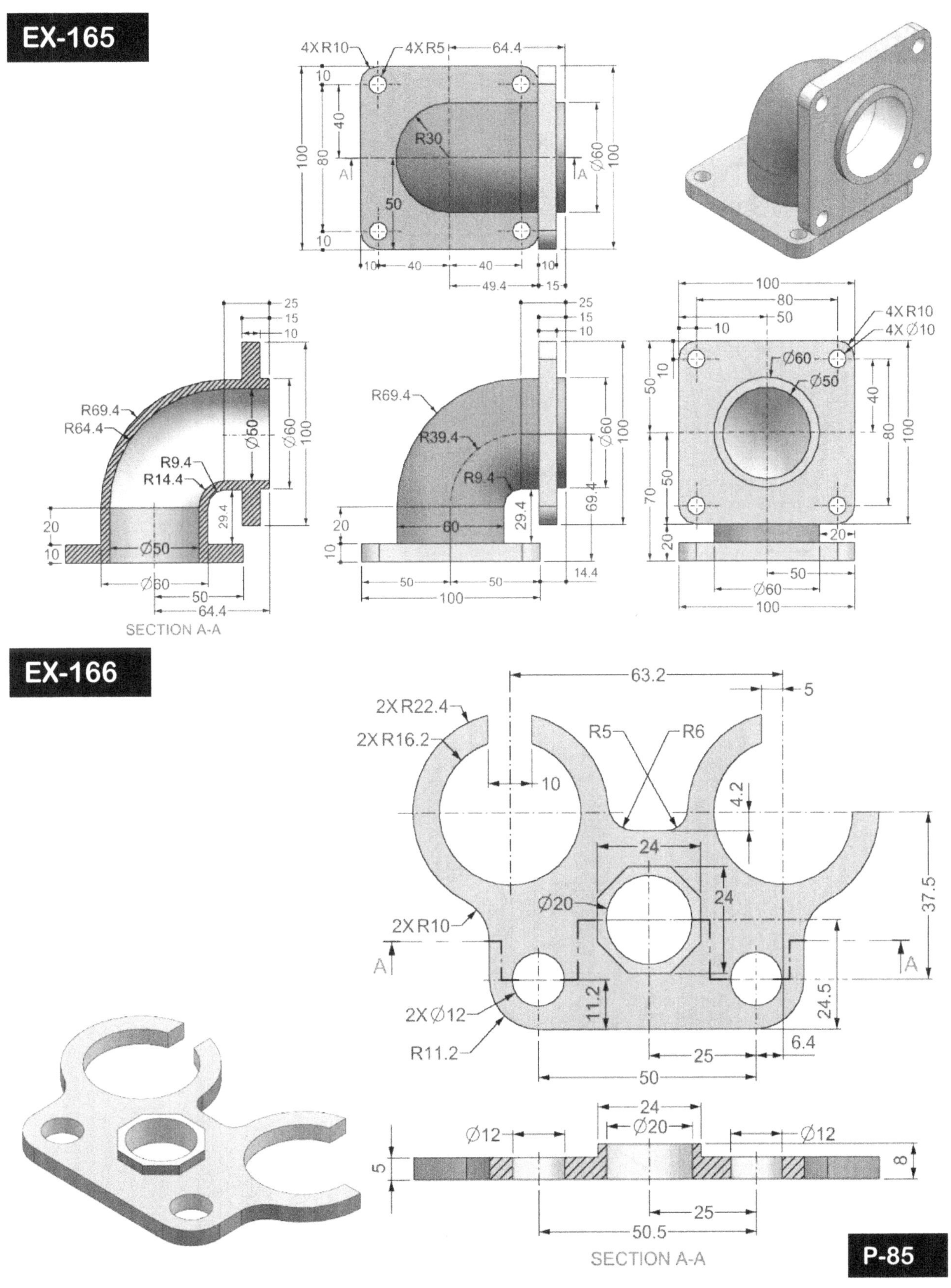
EX-165
4X R10
4X R5
64.4
10
40
80
100
R30
A
A
50
10
Ø60
100
10
40
40
10
49.4
15
25
15
10
R69.4
R64.4
Ø50
Ø60
100
R9.4
R14.4
29.4
20
10
Ø50
Ø60
50
64.4
SECTION A-A
25
15
10
R69.4
R39.4
Ø60
100
R9.4
69.4
20
60
29.4
10
50
50
14.4
100
100
80
50
10
4X R10
4X Ø10
Ø60
Ø50
50
10
40
80
100
70
50
20
20
50
Ø60
100
EX-166
63.2
5
2X R22.4
2X R16.2
R5
R6
10
4.2
24
Ø20
24
2X R10
A
A
2X Ø12
11.2
24.5
R11.2
25
6.4
50
24
Ø12
Ø20
Ø12
5
8
25
50.5
SECTION A-A
P-85

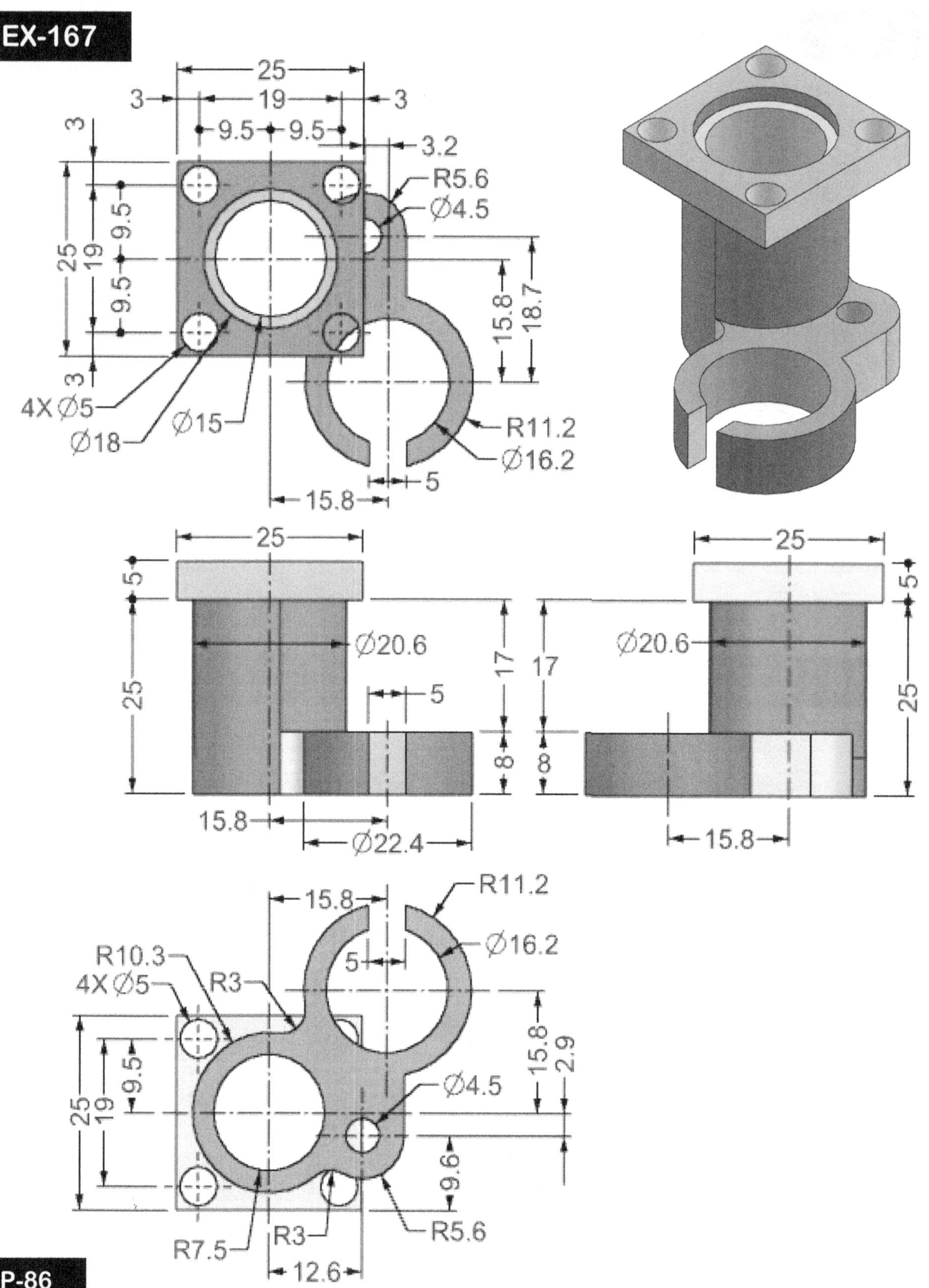

25
3
19
3
9.5
9.5
3.2
R5.6
Ø4.5
25
19
9.5
9.5
9.5
15.8
18.7
4X Ø3
4X Ø5
Ø18
Ø15
R11.2
Ø16.2
5
15.8
25
5
Ø20.6
17
5
25
8
15.8
Ø22.4
25
17
Ø20.6
5
25
8
15.8
15.8
R11.2
R10.3
R3
Ø16.2
5
4X Ø5
15.8
2.9
Ø4.5
25
19
9.5
9.6
R7.5
R3
R5.6
12.6

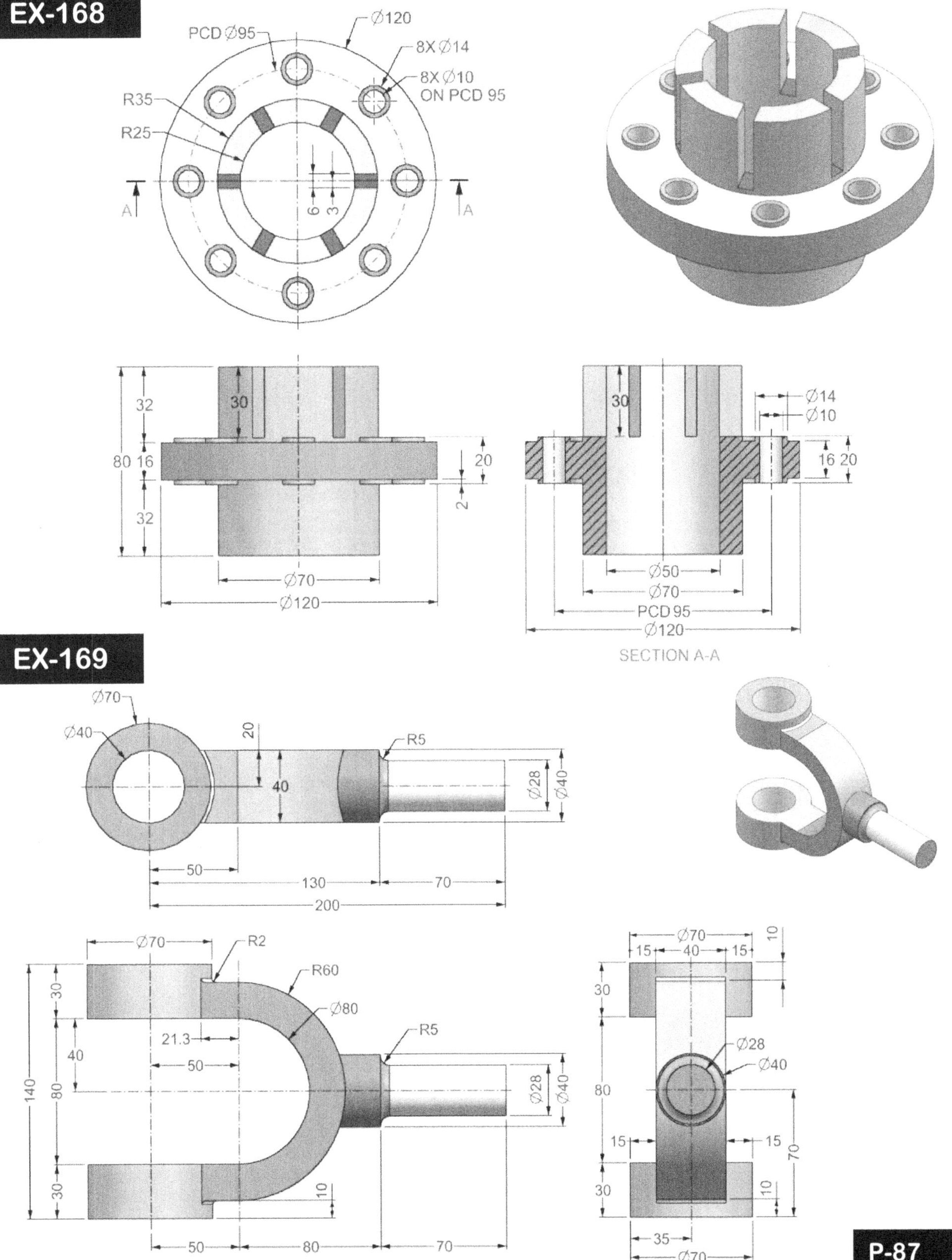

EX-168
PCD Ø95
Ø120
8X Ø14
8X Ø10
ON PCD 95
R35
R25
6
3
A
A
32
30
80 16
20
32
2
Ø70
Ø120
30
Ø14
Ø10
16 20
Ø50
Ø70
PCD 95
Ø120
SECTION A-A
EX-169
Ø70
Ø40
20
R5
40
Ø28
Ø40
50
130
70
200
Ø70
R2
R60
Ø80
R5
30
30
21.3
40
50
80
140
Ø28
Ø40
10
50
80
70
Ø70
15 40 15
10
30
Ø28
Ø40
80
15
15
70
35
30
10
Ø70
P-87

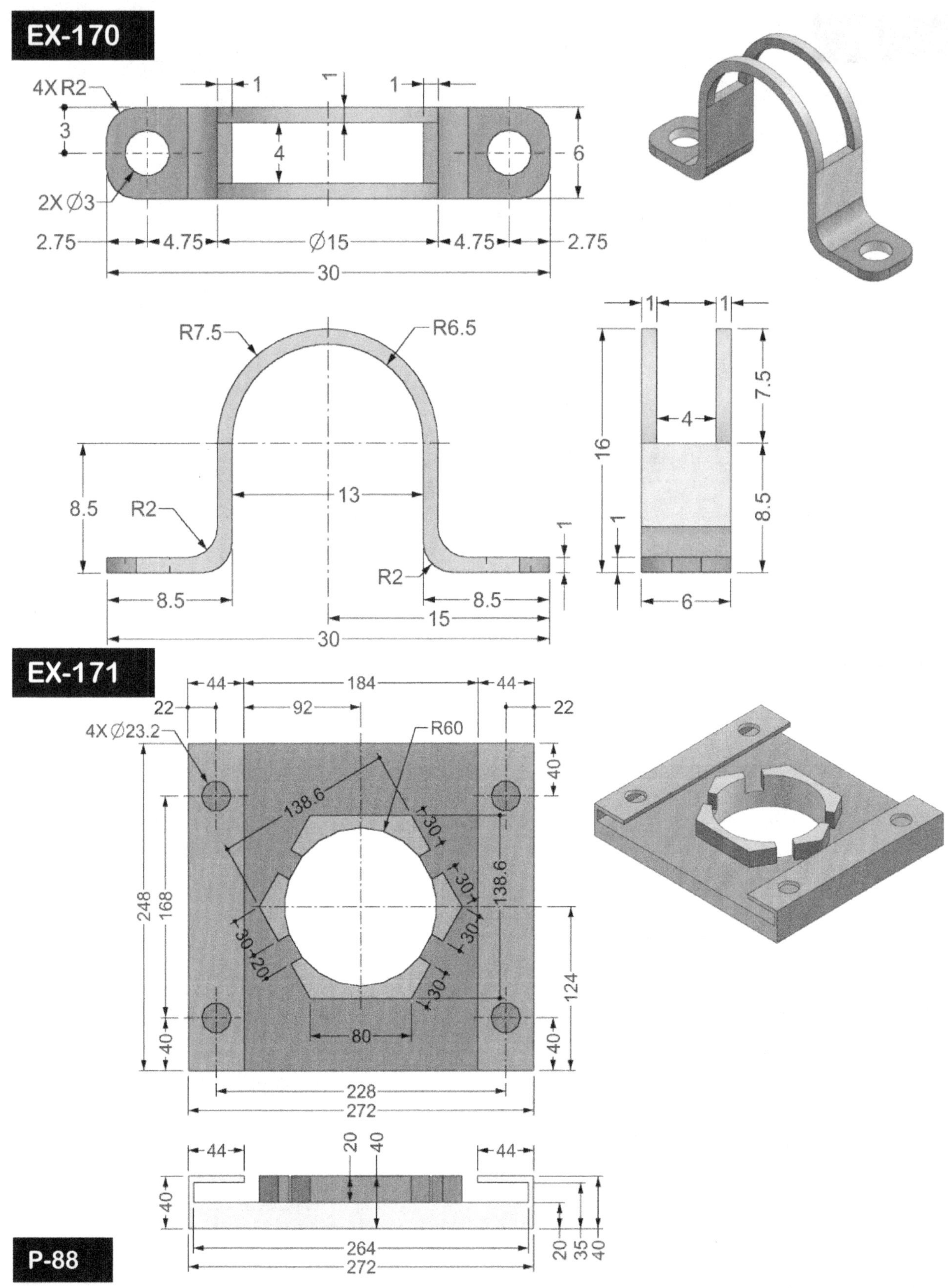

EX-170
4X R2
3
2X Ø3
1
1
1
4
6
2.75
4.75
Ø15
4.75
2.75
30
R7.5
R6.5
R2
R2
13
8.5
8.5
8.5
15
30
1
1
1
7.5
16
4
8.5
1
6
EX-171
44
184
44
22
92
22
4X Ø23.2
R60
40
138.6
30
30
138.6
248
168
30
30
30
20
30
124
80
40
228
272
44
20
40
40
44
264
272
20
35
40
P-88

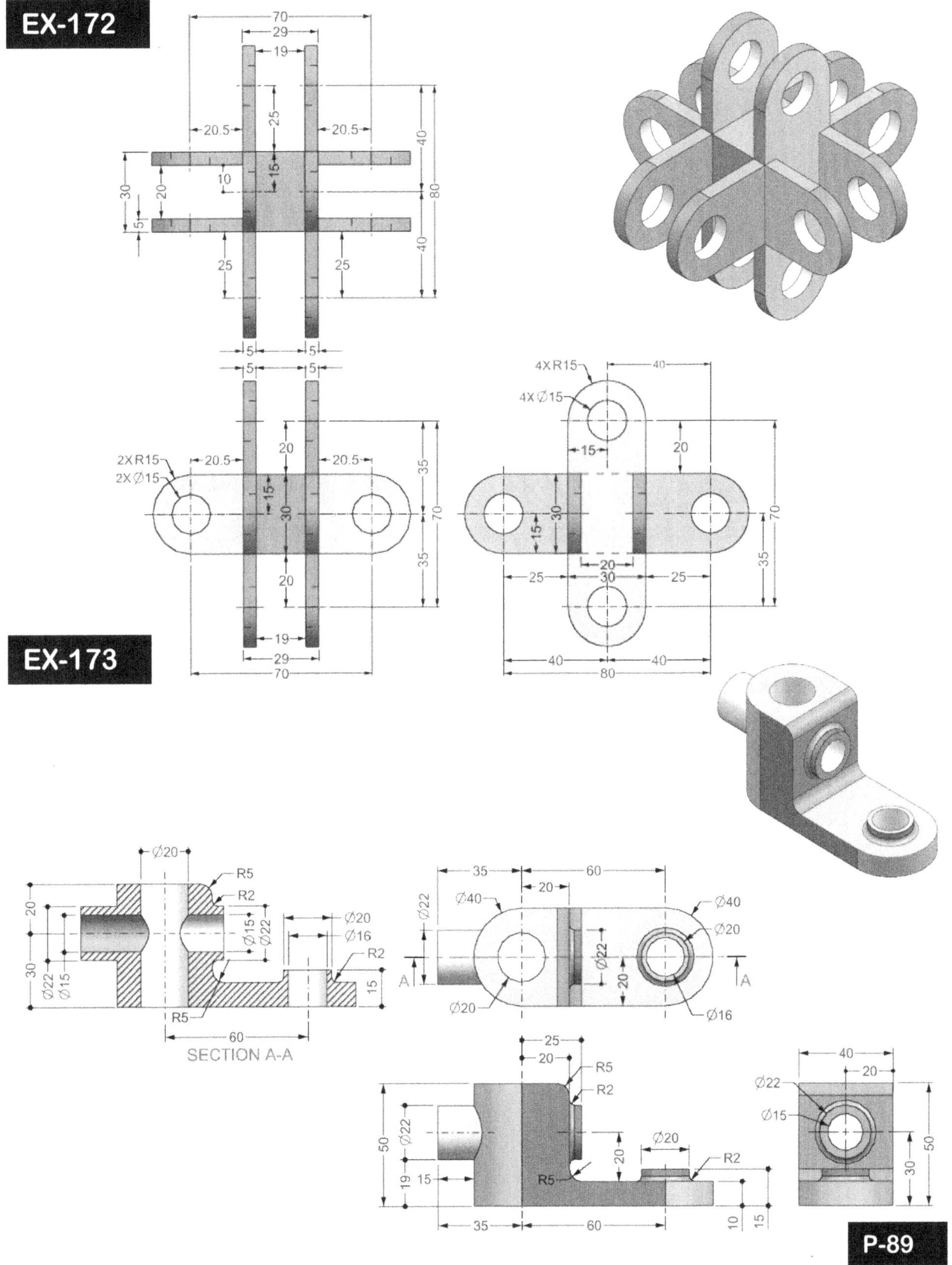

EX-172
EX-173
P-89
SECTION A-A
2X R15
2X Ø15
4X R15
4X Ø15
Ø20
R5
R2
Ø15
Ø22
Ø20
Ø16
R2
R5
Ø22
Ø15
Ø22
Ø40
Ø40
Ø20
Ø22
Ø20
Ø16
Ø22
Ø22
Ø15
R5
R2
Ø20
R2
R5

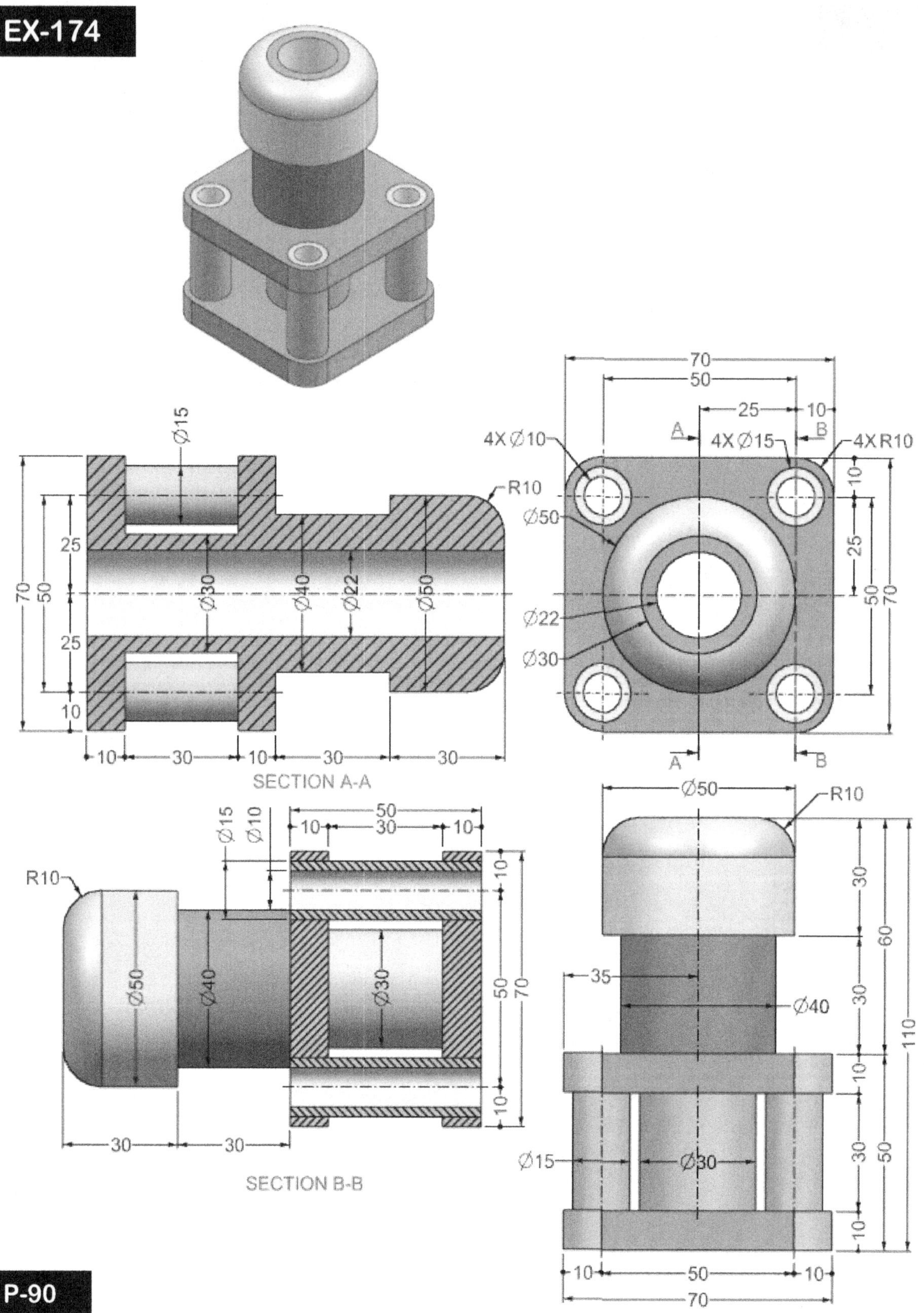

Ø15
R10
25
70
50
25
10
Ø30
Ø40
Ø22
Ø50
10
30
10
30
30
SECTION A-A
Ø15
Ø10
R10
10
30
10
50
Ø50
Ø40
Ø30
50
70
10
30
30
SECTION B-B
70
50
25
10
4X Ø10
A
4X Ø15
B
4X R10
R10
Ø50
Ø22
Ø30
10
25
50
70
A
B
Ø50
R10
30
60
35
Ø40
30
110
10
30
50
Ø15
Ø30
10
10
50
10
70

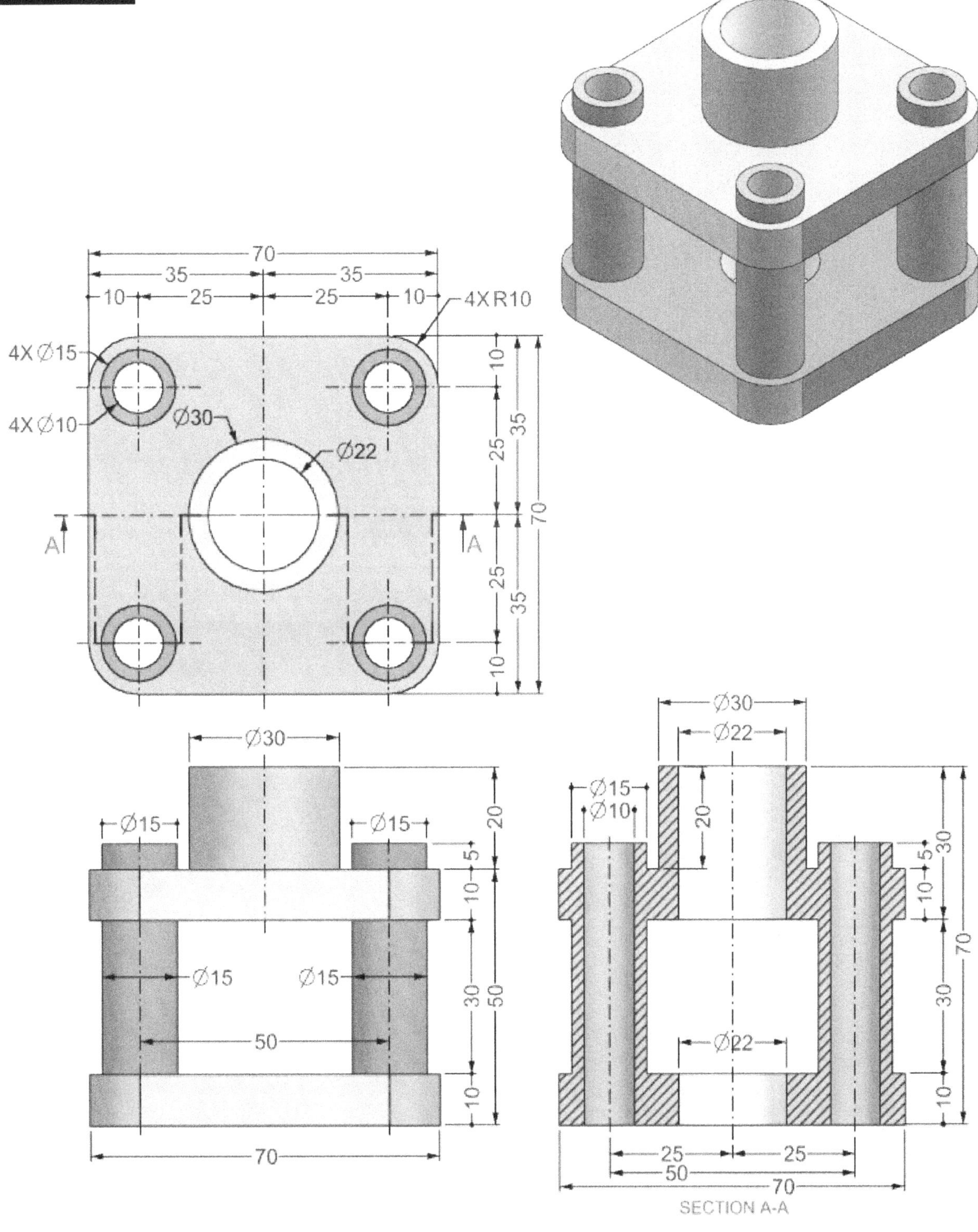

70
35
35
10
25
25
10
4X R10
4X Ø15
4X Ø10
Ø30
Ø22
10
25
35
70
A
A
25
35
10
Ø30
Ø15
Ø15
20
5
10
Ø15
Ø15
30
50
50
10
70
Ø30
Ø22
Ø15
Ø10
20
5
10
30
70
Ø22
30
25
25
50
70
SECTION A-A

EX-176
65 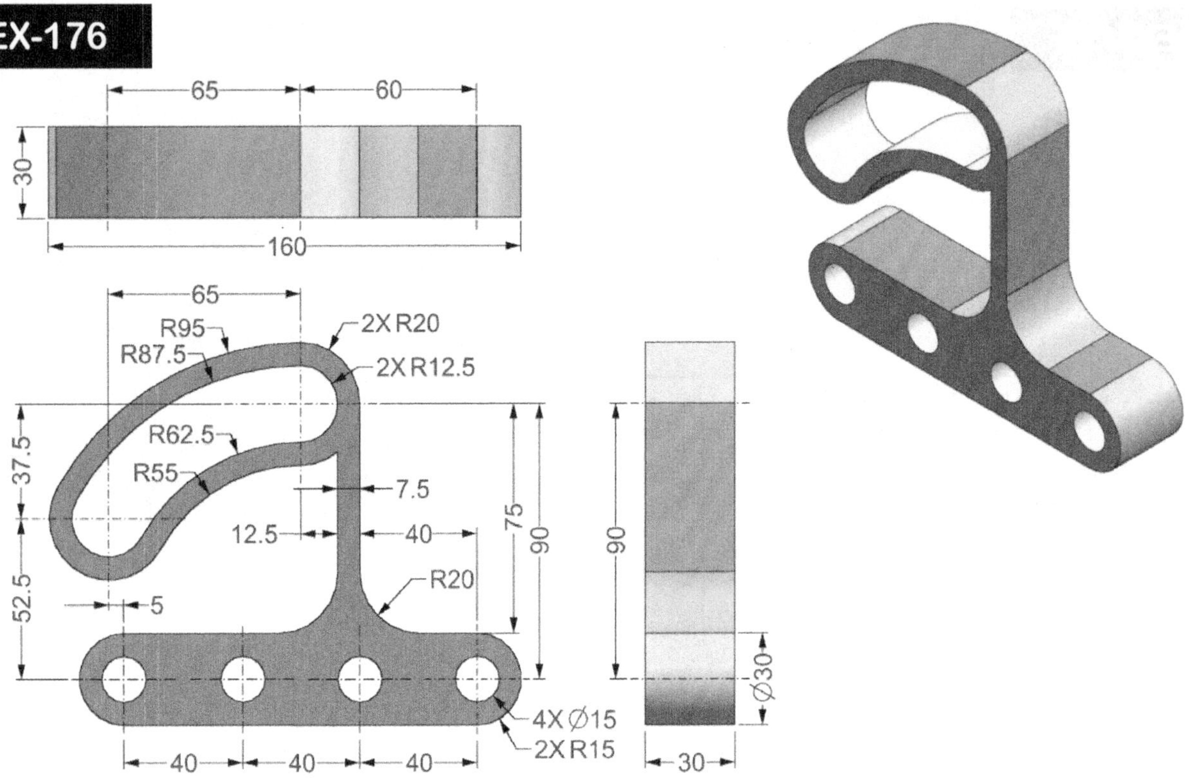
60
30
160
65
R95
R87.5
2X R20
2X R12.5
R62.5
R55
37.5
7.5
12.5
40
75
90
90
R20
52.5
5
Ø30
4X Ø15
2X R15
40
40
40
30
EX-177
R20 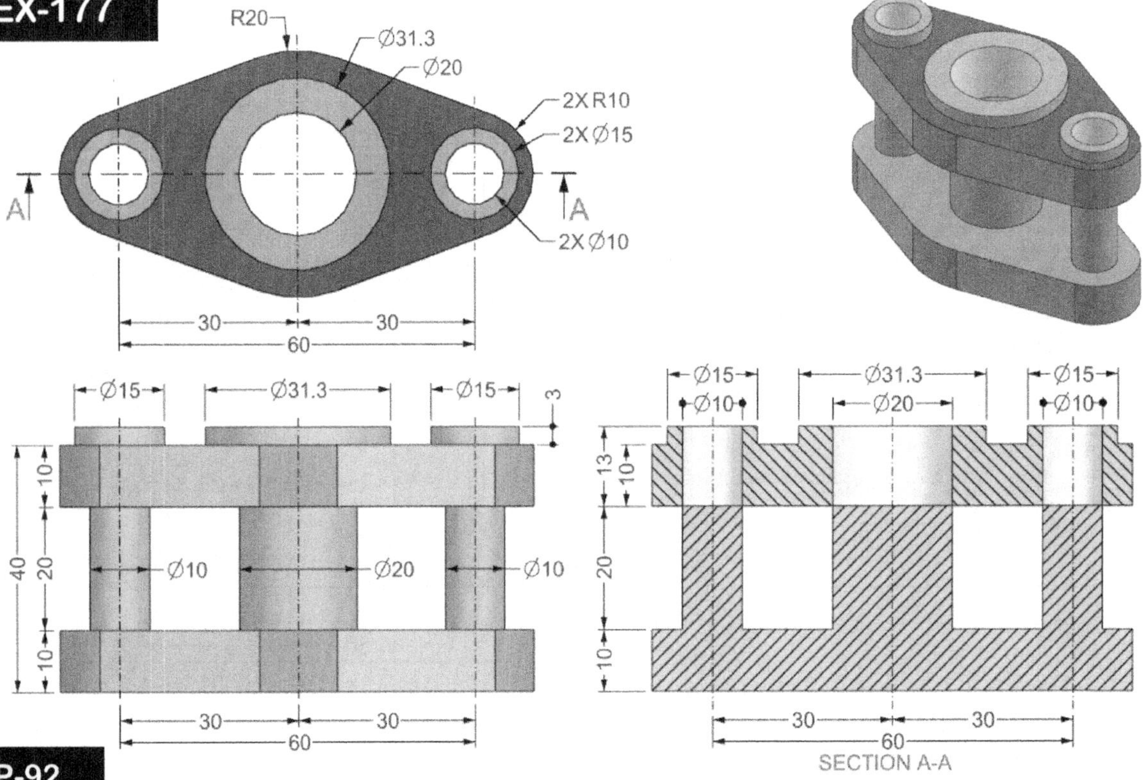
Ø31.3
Ø20
2X R10
2X Ø15
2X Ø10
A
A
30
30
60
Ø15
Ø31.3
Ø15
3
10
40
20
Ø10
Ø20
Ø10
10
30
30
60
Ø15
Ø31.3
Ø15
Ø10
Ø20
Ø10
13
10
20
10
30
30
60
SECTION A-A
P-92

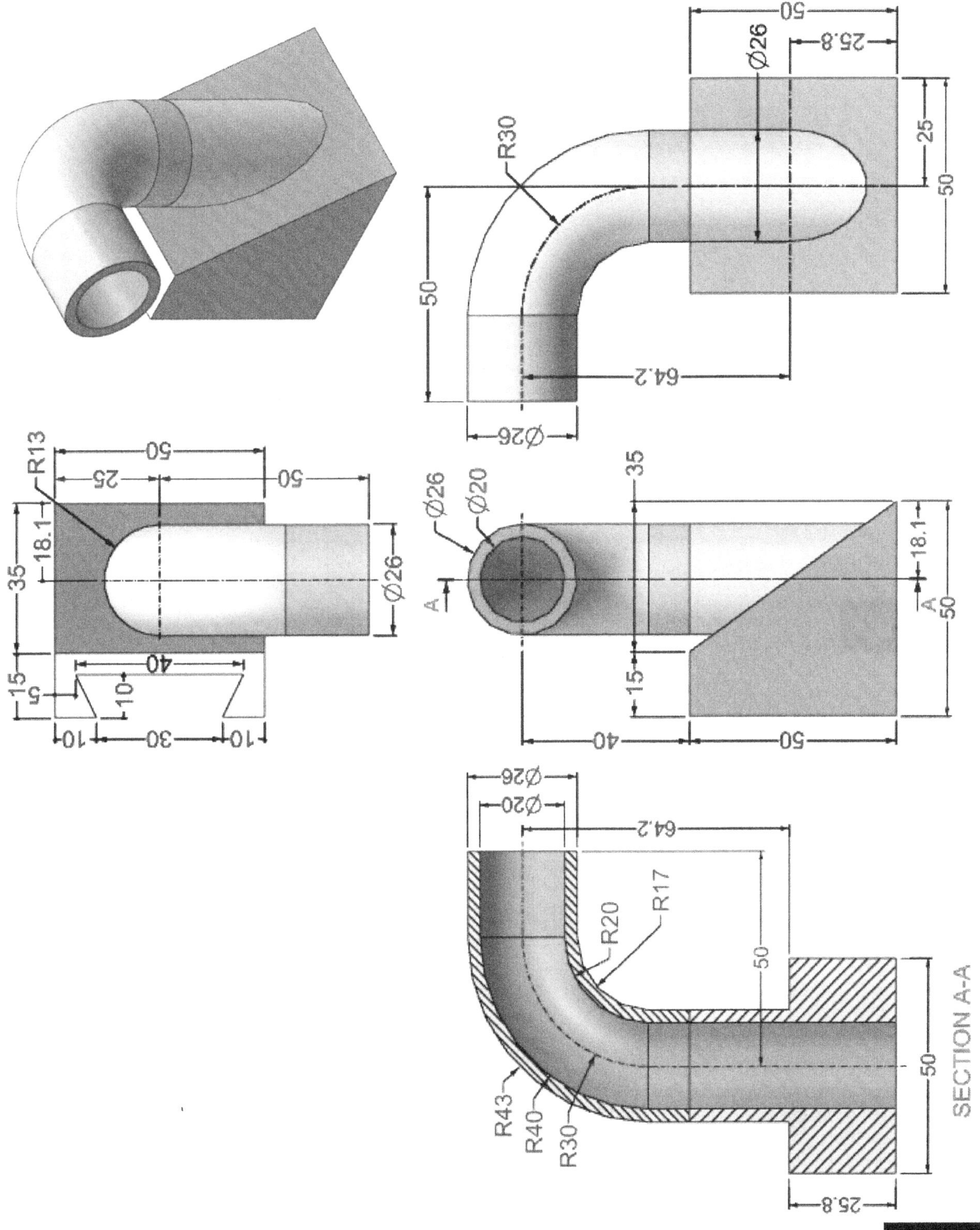

Ø26
50
25.8
R30
25
50
64.2
Ø26
R13
50
25
50
Ø26
35
18.1
40
5
10
10
30
10
15
Ø26
Ø20
A
35
A
18.1
50
15
40
50
Ø26
Ø20
64.2
R20
R17
50
R43
R40
R30
50
25.8
SECTION A-A

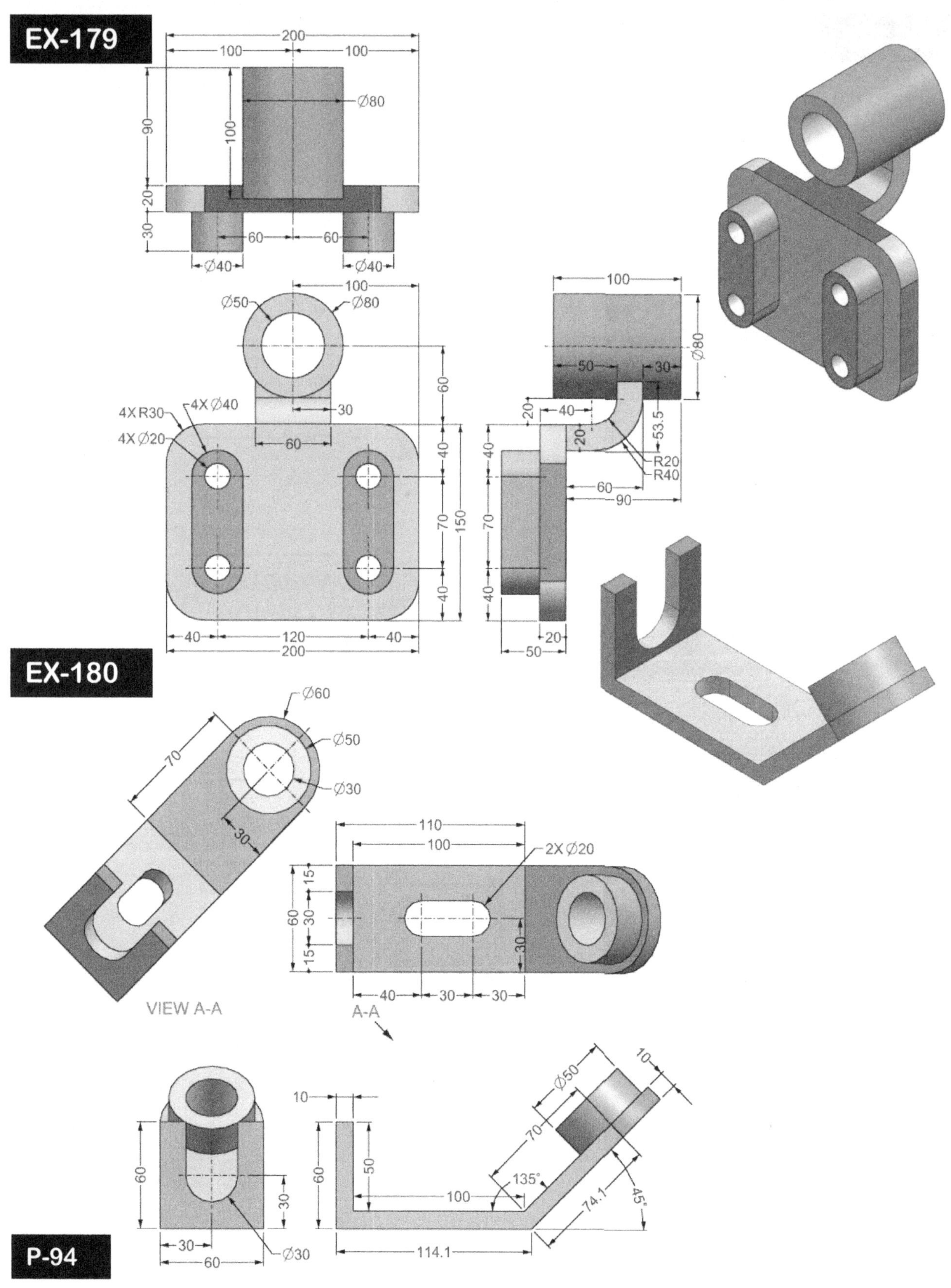

EX-179
EX-180
P-94
Ø80
200
100
100
90
100
20
30
60
60
Ø40
Ø40
Ø50
Ø80
4X R30
4X Ø40
4X Ø20
60
30
60
40
70
150
40
40
120
40
200
100
50
30
Ø80
20
40
20
53.5
R20
R40
60
90
40
70
40
20
50
Ø60
Ø50
70
Ø30
30
110
100
2X Ø20
15
30
60
30
15
40
30
30
VIEW A-A
A-A
60
30
30
60
Ø30
10
60
50
100
Ø50
10
70
135°
45°
74.1
114.1

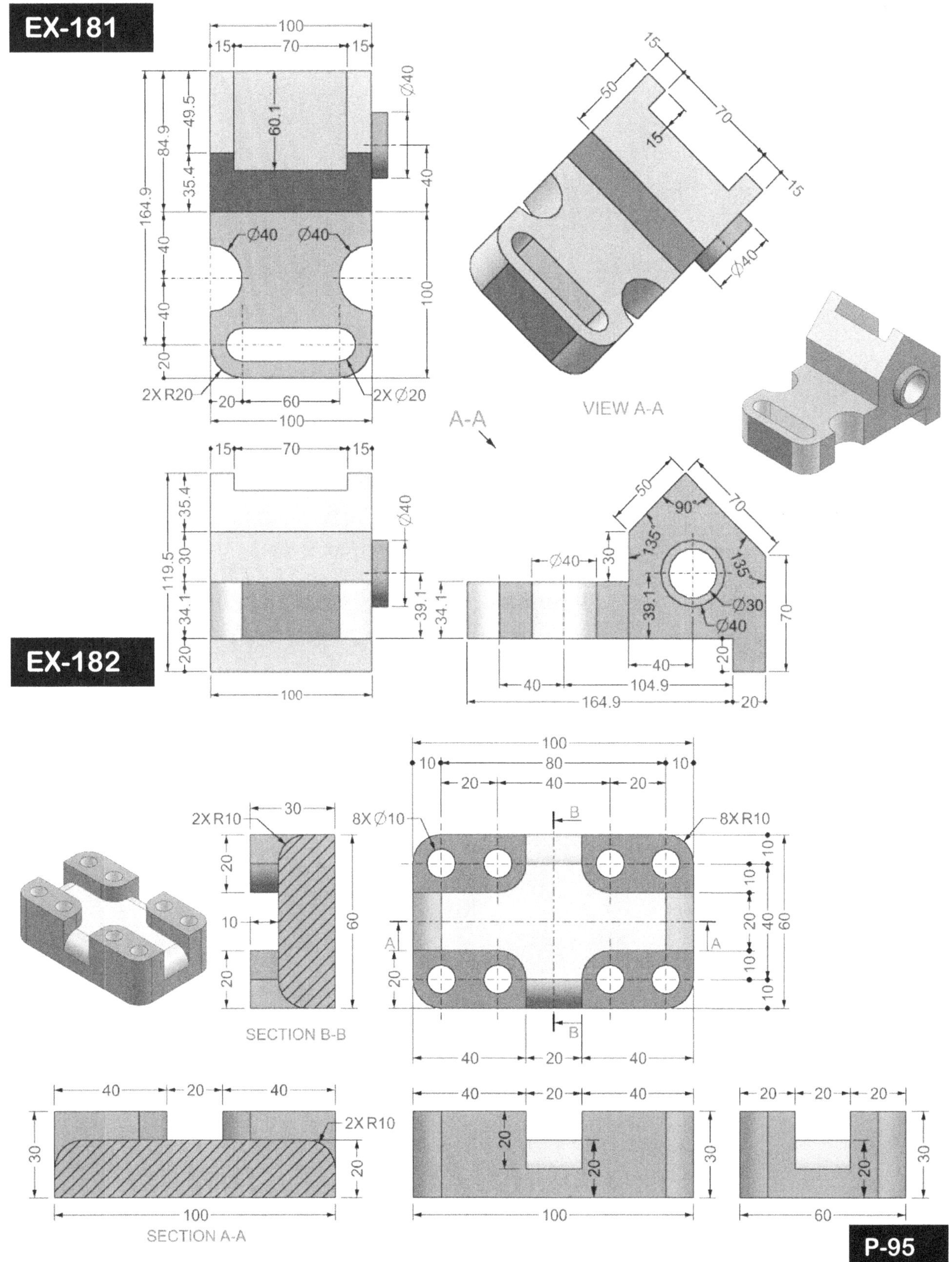

EX-181
100
15 70 15
49.5
60.1
84.9
35.4
164.9
Ø40
40
40
Ø40 Ø40
100
40
20
2X R20
20 60
2X Ø20
100
15 70 15
50
15
70
15
Ø40
VIEW A-A
A-A
EX-182
15 70 15
35.4
30
119.5
34.1
20
Ø40
Ø40
39.1
100
50
90°
135
135
30
Ø40
34.1
39.1
Ø30
Ø40
40
40
104.9
20
164.9
20
70
100
80
10 10
20 40 20
B
8X Ø10
8X R10
10
10
20 40 60
2X R10
30
20
10
60
20
A A
20
SECTION B-B
40 20 40
B
40 20 40
40 20 40
20 20 20
30
20
2X R10
30
20
20
30
100
SECTION A-A
100
60
P-95

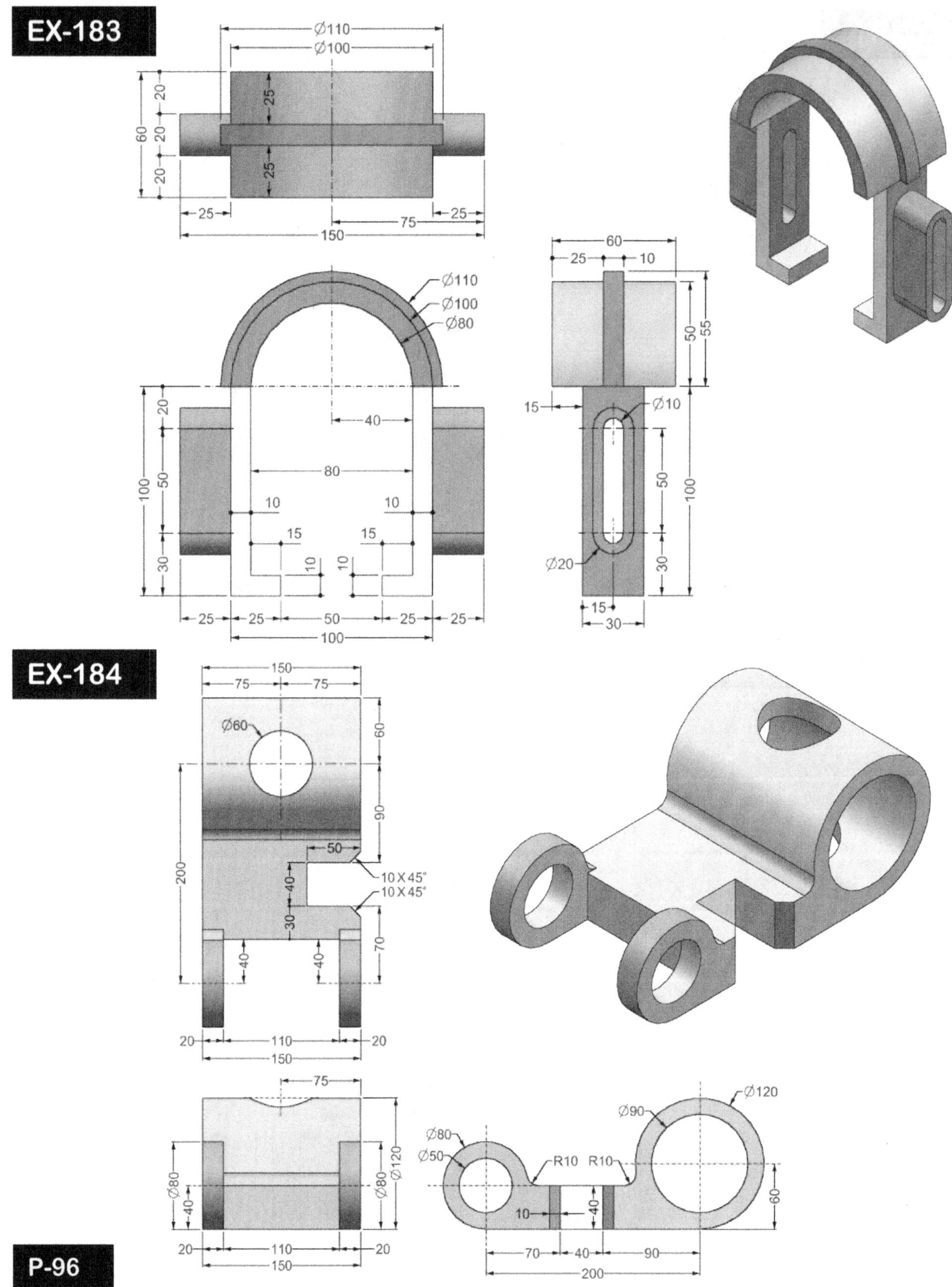

EX-183
Ø110
Ø100
20
20
60
20
25
25
25
75
25
150
Ø110
Ø100
Ø80
20
40
50
100
80
10
10
15
15
10
10
30
25
25
50
25
25
100
60
25
10
50
55
15
Ø10
50
100
30
Ø20
15
30
EX-184
150
75
75
Ø60
60
90
200
50
10 X 45°
40
10 X 45°
30
40
40
70
20
110
20
150
75
Ø80
Ø120
Ø80
40
20
110
20
150
Ø80
Ø90
Ø120
Ø50
R10
R10
10
40
60
70
40
90
200
P-96

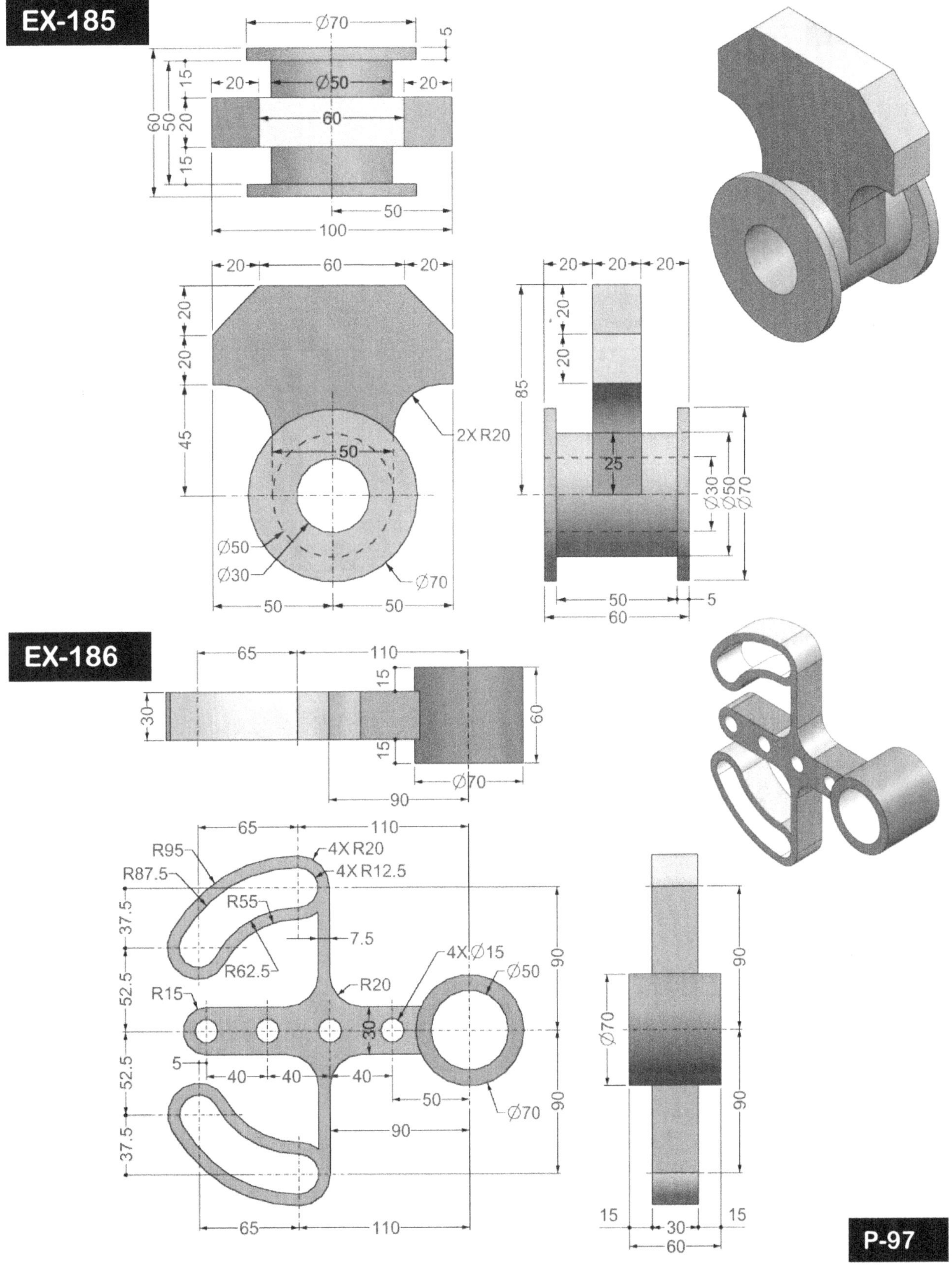

EX-185
EX-186
P-97

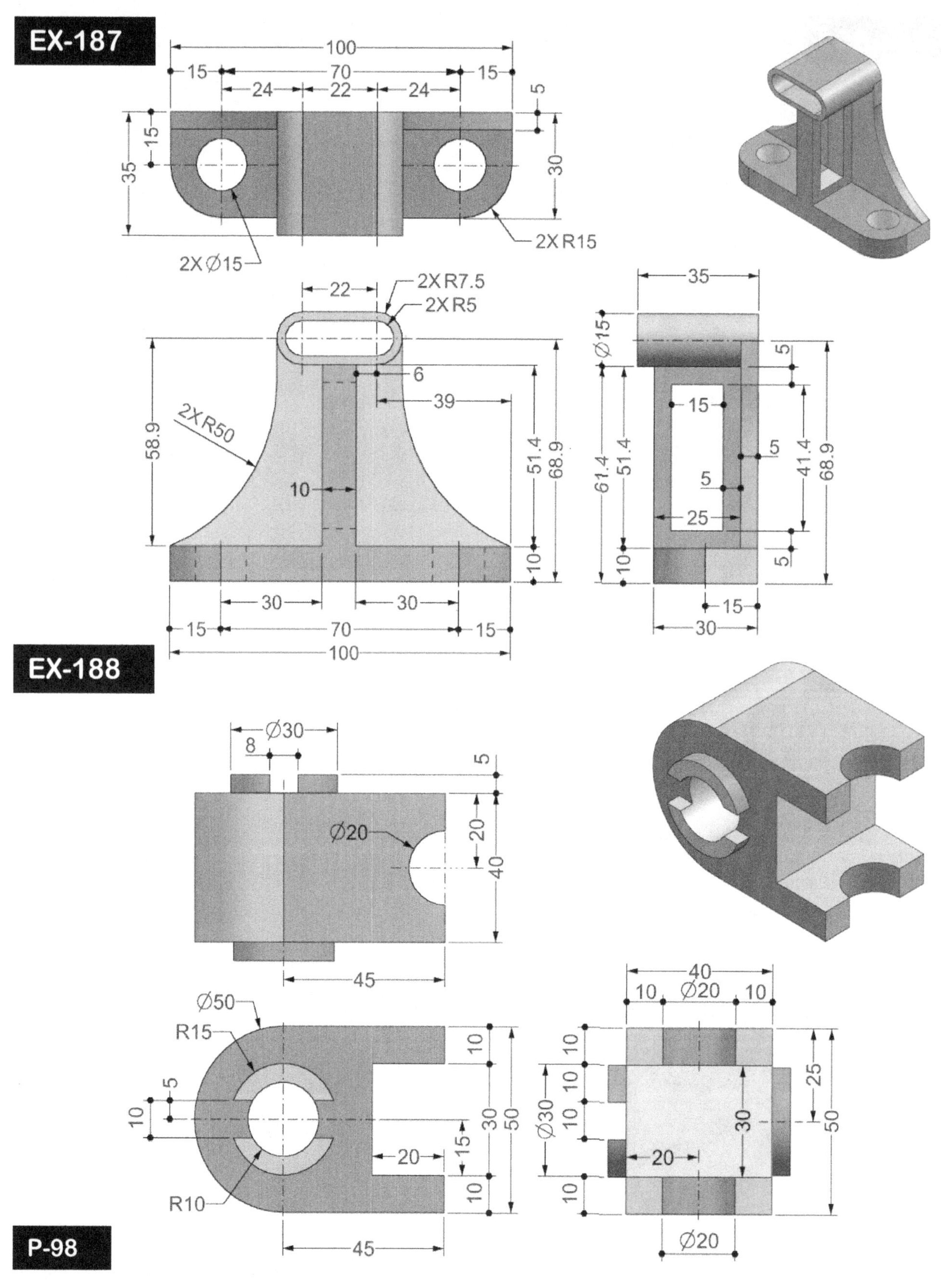

EX-187
100
15
70
15
24
22
24
5
15
35
30
2X Ø15
2X R15
22
2X R7.5
2X R5
6
39
2X R50
58.9
10
51.4
68.9
30
30
10
15
70
15
100
35
Ø15
5
15
5
61.4
51.4
5
5
25
41.4
68.9
10
5
15
30
EX-188
Ø30
8
5
Ø20
20
40
45
Ø50
R15
5
10
10
30
50
Ø20
15
20
R10
45
40
10
Ø20
10
10
10
Ø30
30
25
10
50
20
10
Ø20
P-98

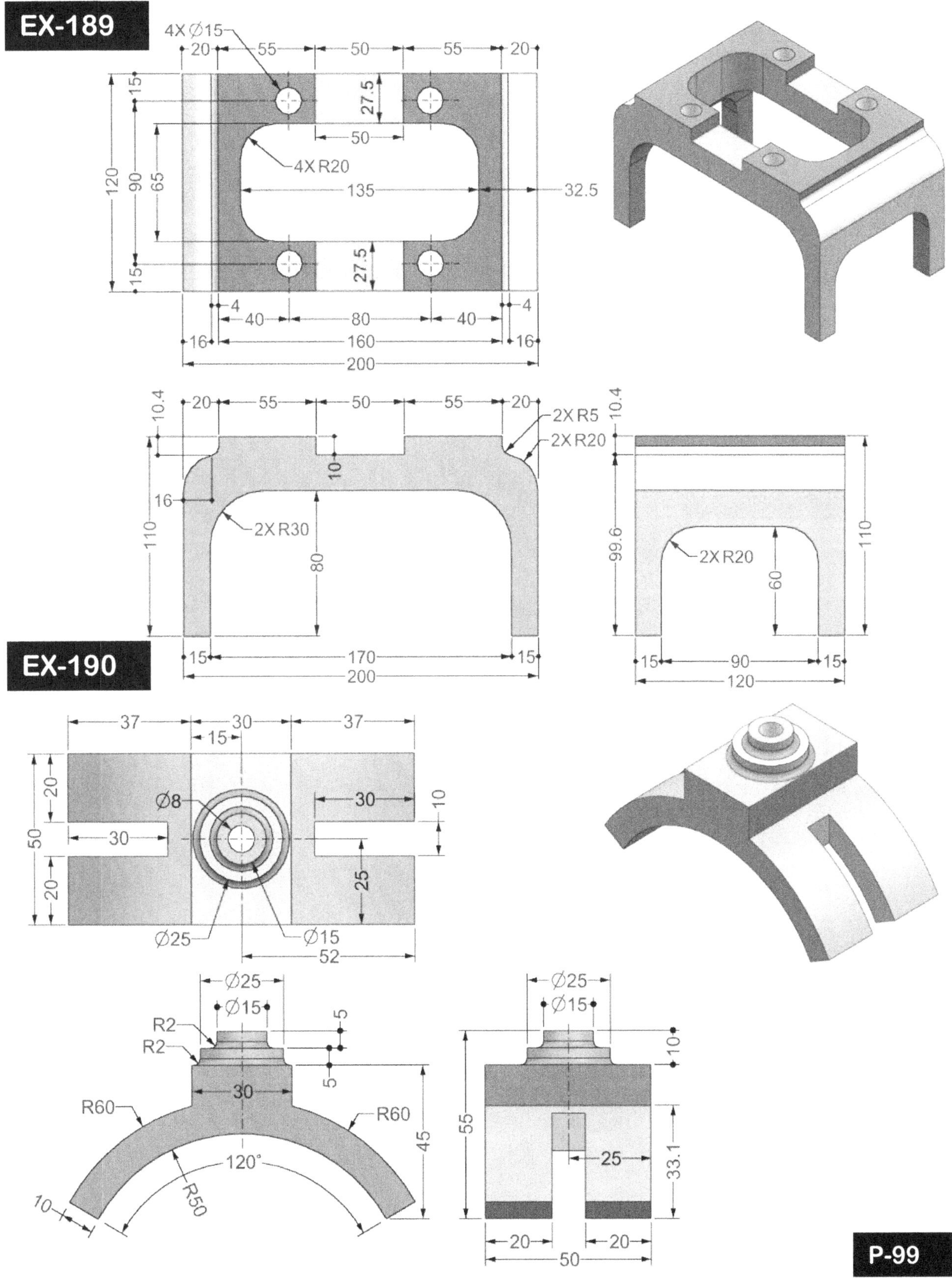

EX-189
4X Ø15
20
55
50
55
20
15
27.5
120
90
65
50
4X R20
135
32.5
15
27.5
4
40
80
40
4
16
160
16
200

EX-190
10.4
20
55
50
55
20
2X R5
2X R20
10.4
16
10
2X R30
110
80
99.6
2X R20
110
60
15
170
15
15
90
15
200
120

37
30
37
15
20
Ø8
30
10
50
30
30
20
25
Ø25
Ø15
52

Ø25
Ø15
R2
R2
5
5
R60
30
R60
45
R60
R60
120°
R50
10

Ø25
Ø15
10
55
25
33.1
20
20
50

P-99

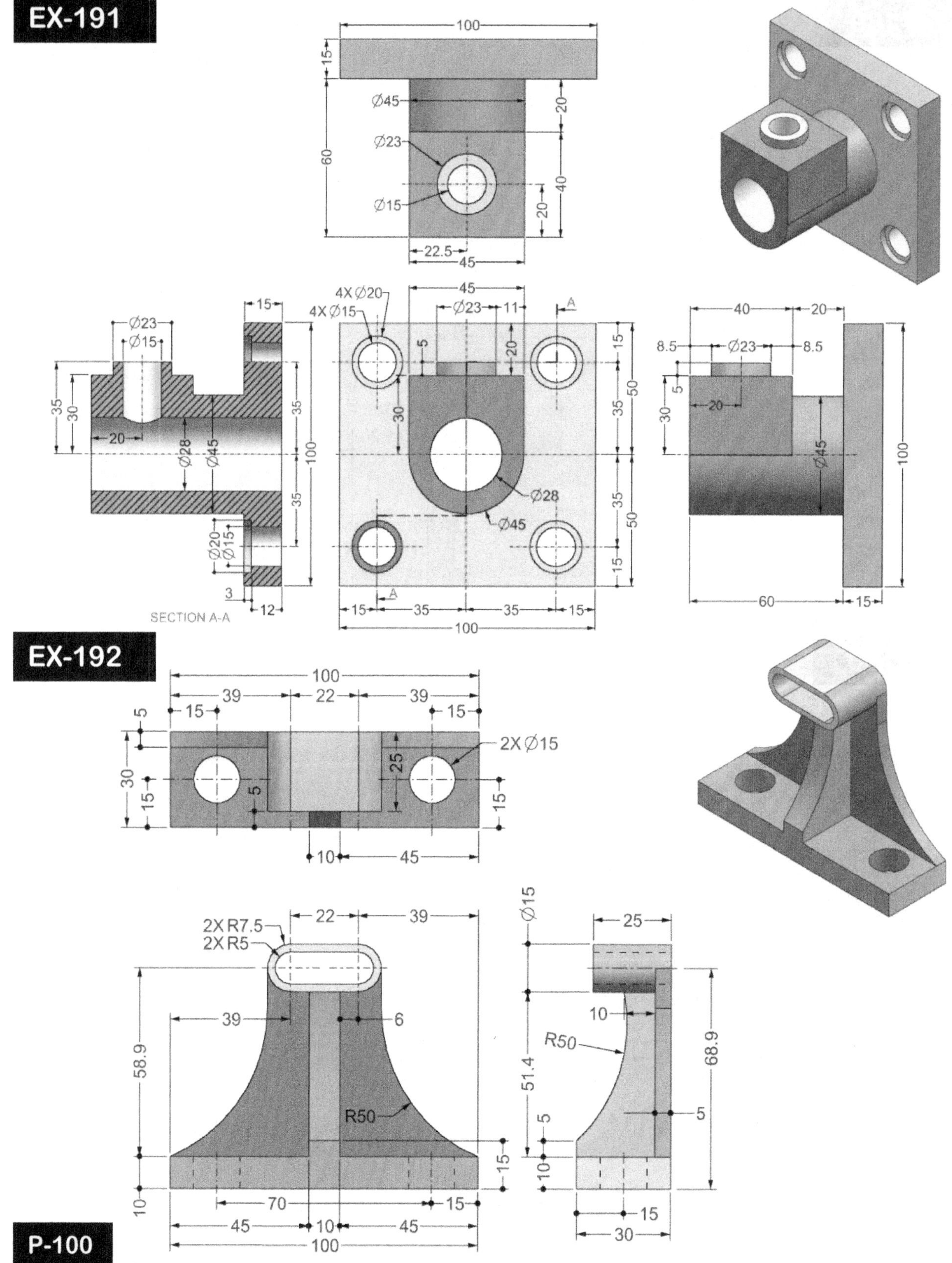

EX-191
100
15
Ø45
Ø23
Ø15
60
20
40
20
22.5
45
4X Ø20
4X Ø15
45
Ø23
11
A
15
5
20
35
50
35
50
15
Ø23
Ø15
15
35
30
20
Ø28
Ø45
35
Ø20
Ø15
3
12
100
A
SECTION A-A
Ø28
Ø45
15
35
35
15
35
35
15
100
40
20
8.5
Ø23
8.5
5
30
20
Ø45
100
60
15
EX-192
100
39
22
39
5
15
15
30
25
2X Ø15
15
15
10
45
58.9
2X R7.5
2X R5
22
39
Ø15
39
6
R50
70
15
45
10
45
100
10
25
10
R50
51.4
68.9
5
10
5
15
30
P-100

EX-193

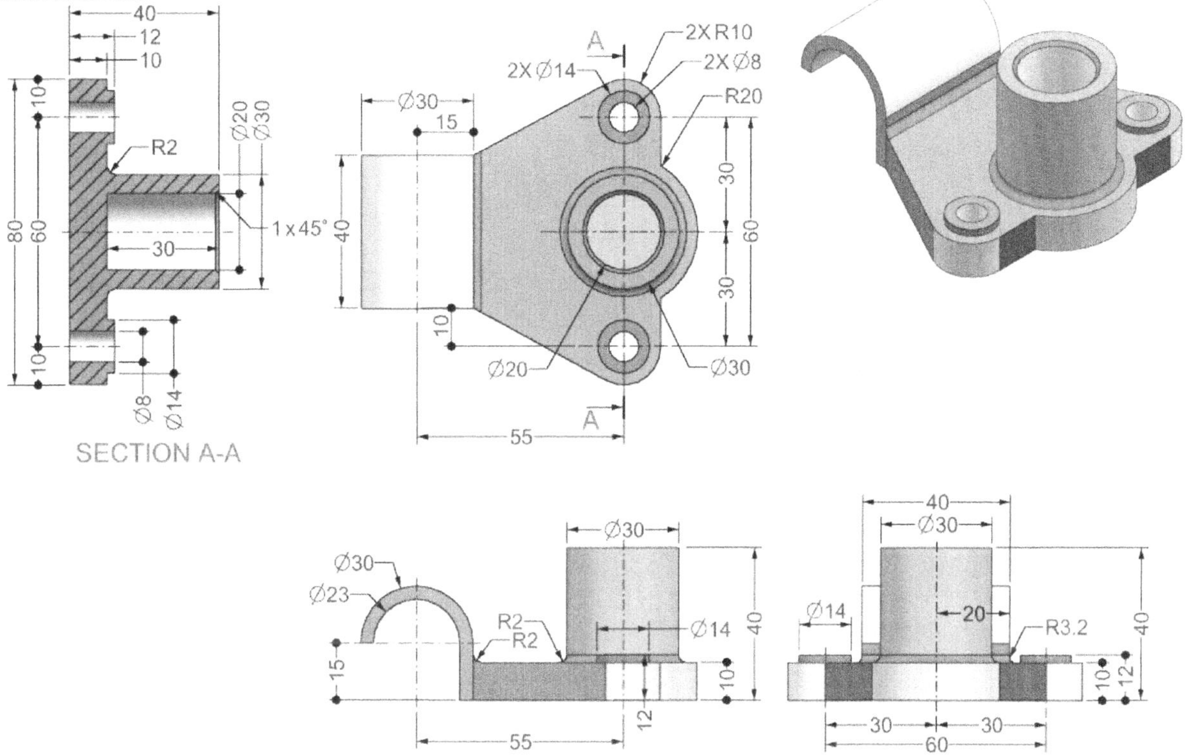

SECTION A-A

EX-194

ALL HOLES CHAMFER 2MM

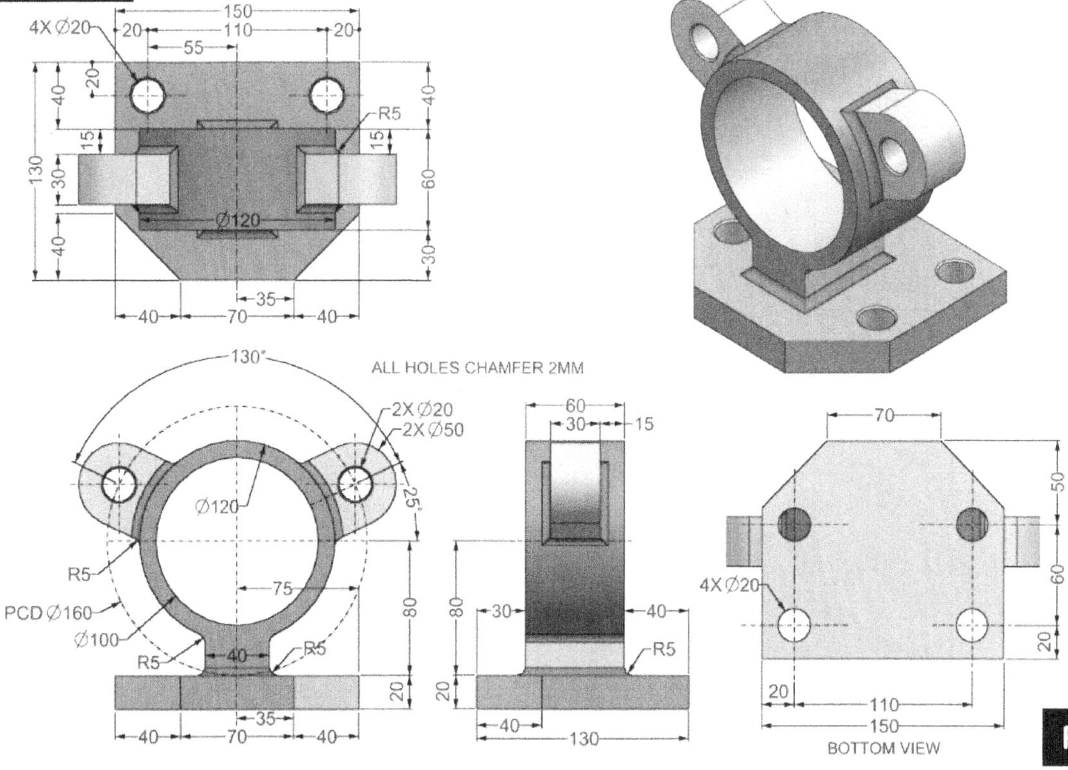

BOTTOM VIEW

P-101

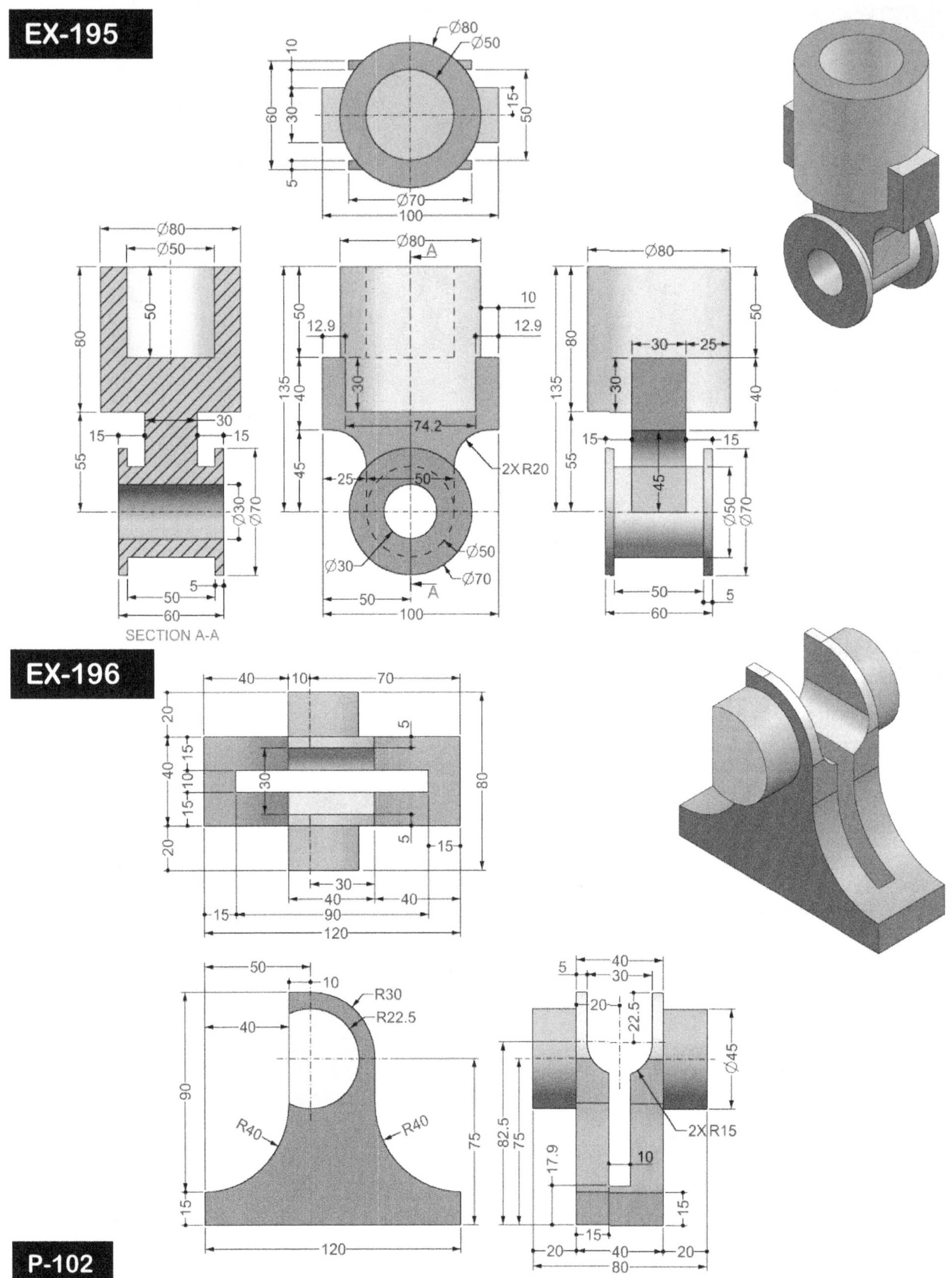

EX-195
SECTION A-A
EX-196
P-102

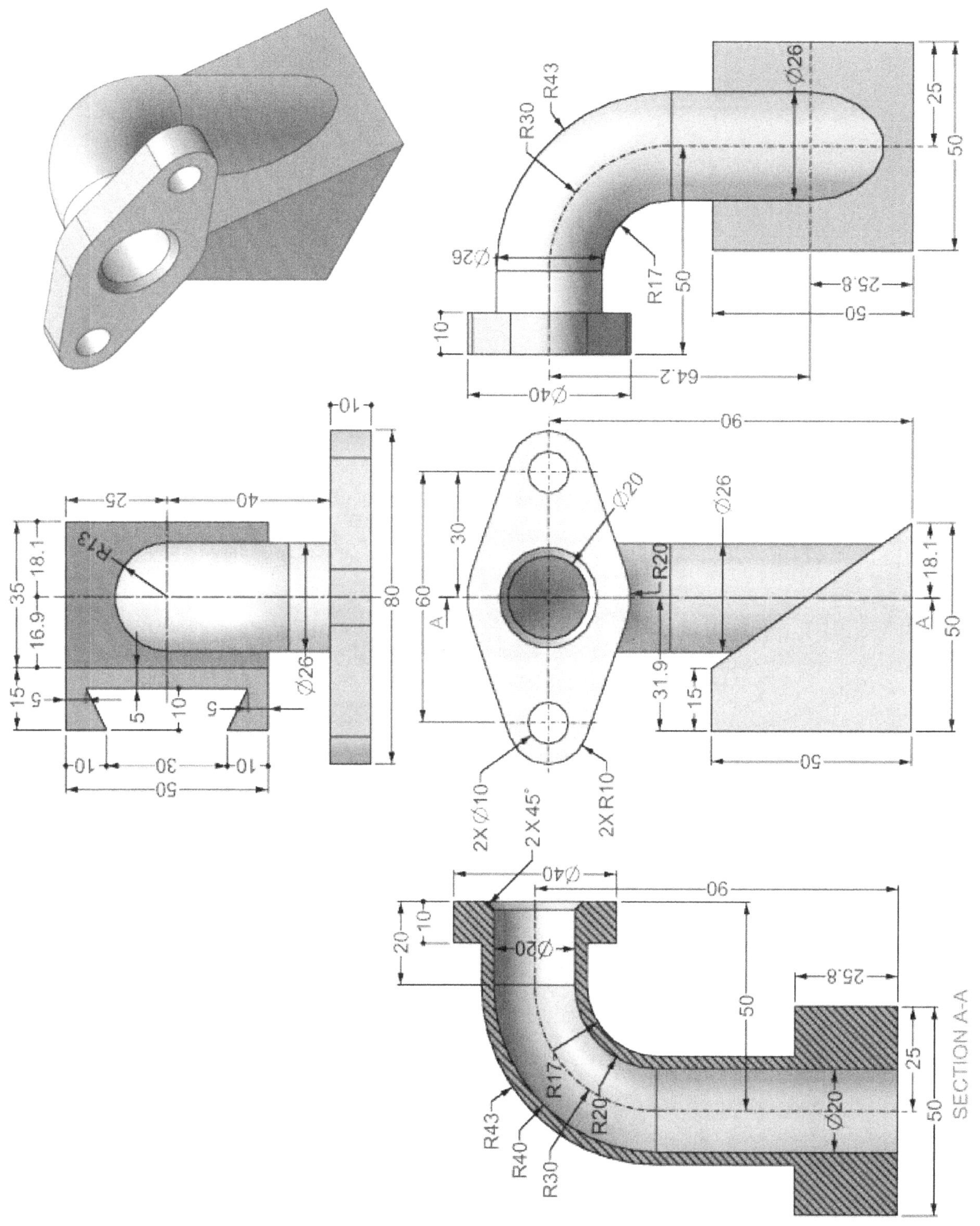

R30
R43
Ø26
25
50
Ø26
R17
50
25.8
50
10
64.2
Ø40
10
25
40
R13
35
18.1
16.9
15
5
5
10
5
Ø26
80
10
30
30
60
A
A
Ø20
Ø26
R20
31.9
15
18.1
50
50
90
2X Ø10
2X 45°
2X R10
20
10
Ø40
Ø20
90
25.8
R17
R20
50
25
Ø20
50
R43
R40
R30
SECTION A-A

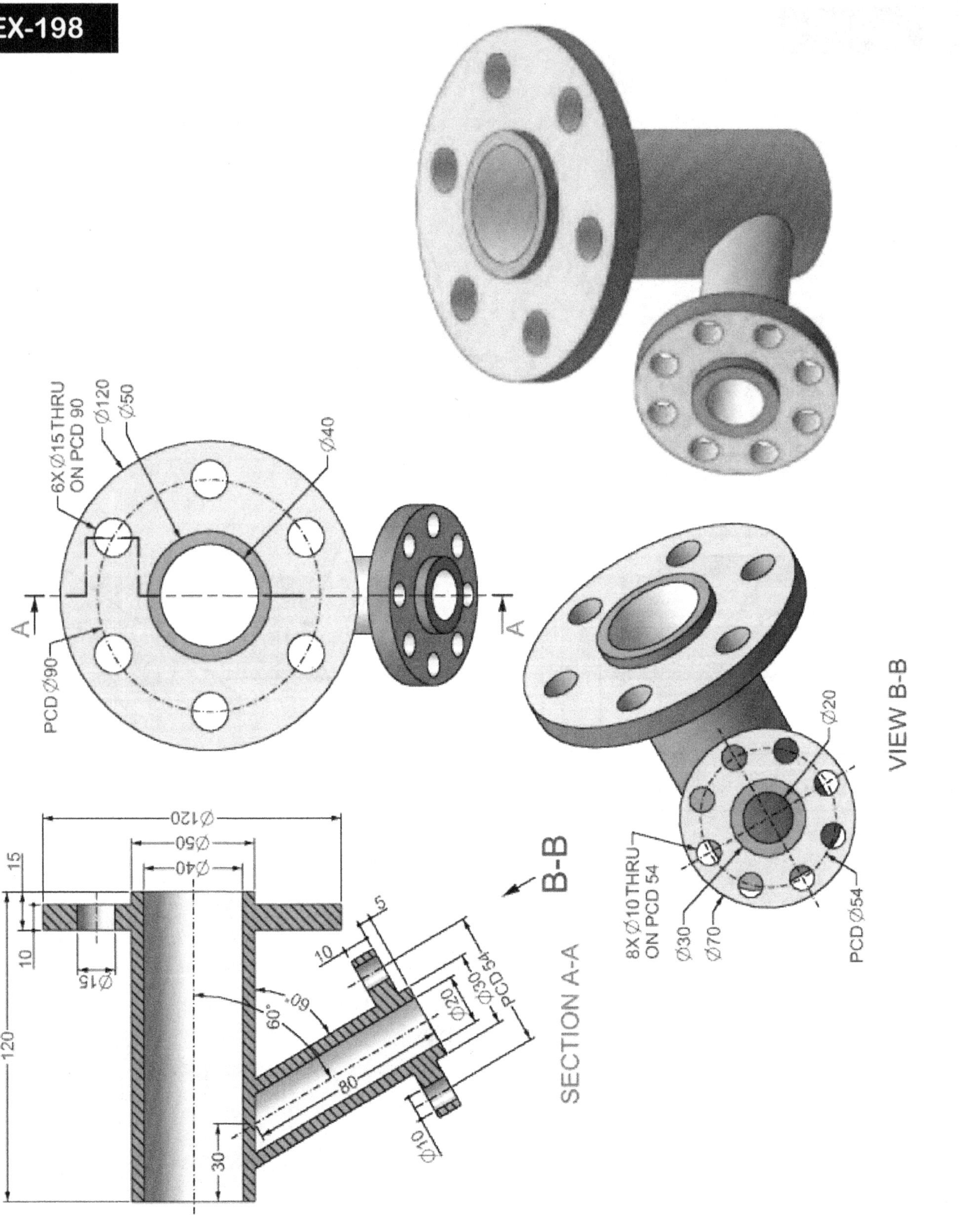

6X Ø15 THRU
ON PCD 90
Ø120
Ø50
Ø40
PCD Ø90
A
A
Ø120
Ø50
Ø40
15
10
Ø15
120
60°
60°
80
30
5
10
Ø20
Ø30
Ø54
PCD 54
Ø10
SECTION A-A
B-B
VIEW B-B
8X Ø10 THRU
ON PCD 54
Ø20
Ø30
Ø70
PCD Ø54

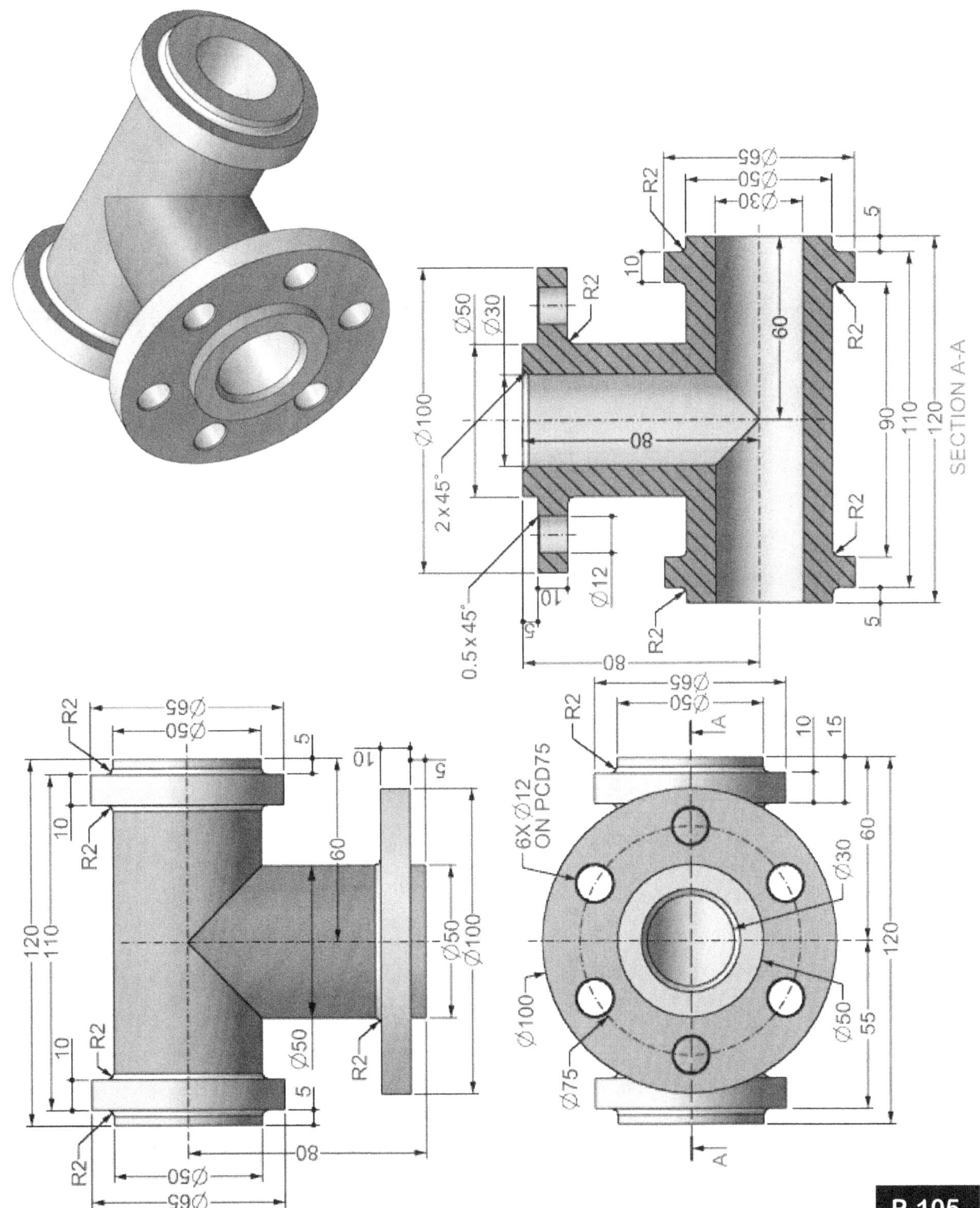

Ø65
Ø50
Ø30
R2
10
60
R2
5
90
110
120
SECTION A-A
Ø100
Ø50
Ø30
R2
80
2 x 45°
0.5 x 45°
R2
10
Ø12
5
80
Ø65
Ø50
R2
R2
5
10
10
60
120
110
R2
R2
10
5
80
Ø50
Ø65
Ø50
Ø100
Ø50
R2
Ø65
Ø50
R2
IA
10
15
6X Ø12
ON PCD75
Ø30
60
120
Ø100
Ø50
55
Ø75
A

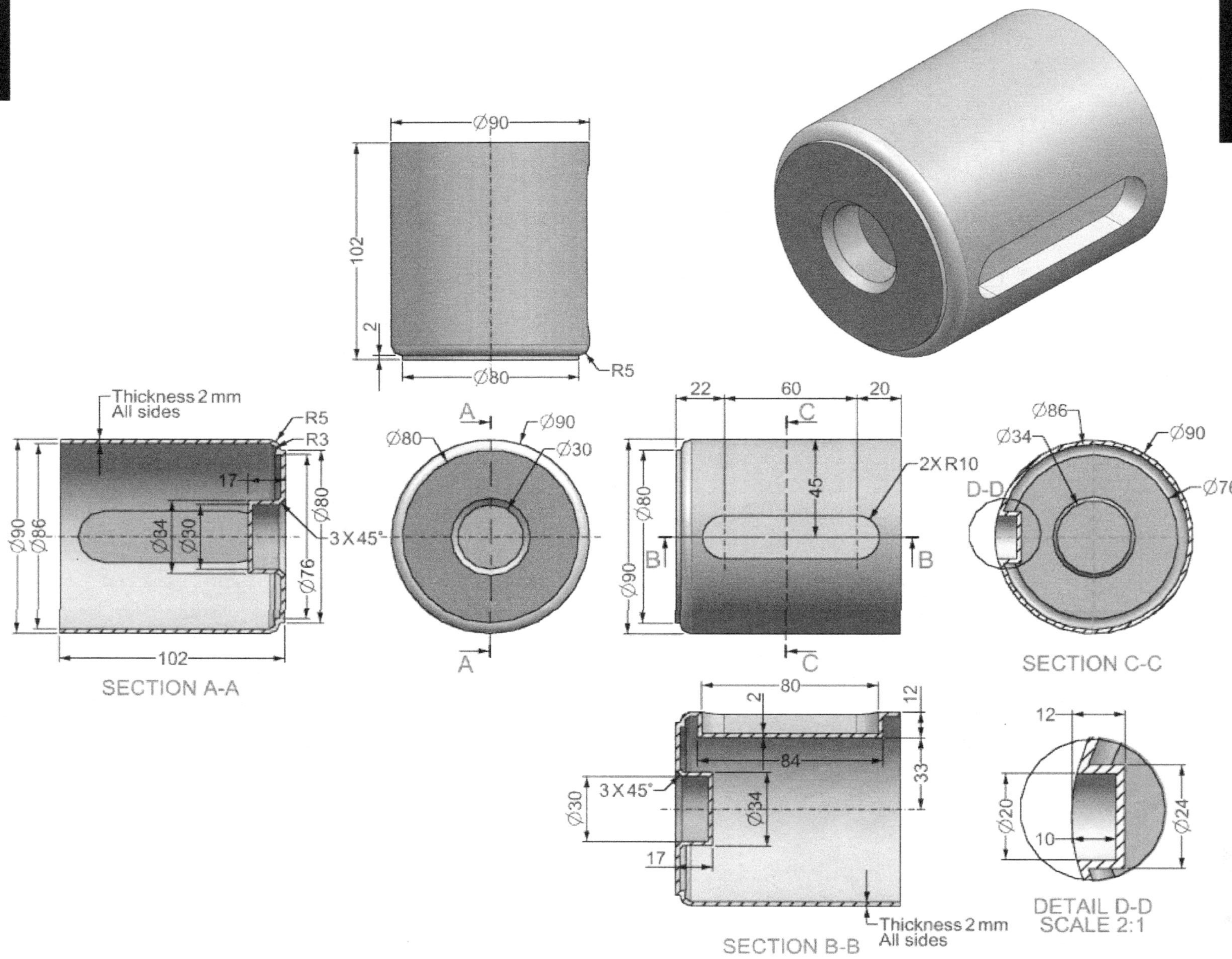

P-106
EX-200
Ø90
102
2
Ø80
R5
Thickness 2 mm
All sides
R5
R3
17
Ø80
Ø90
Ø86
Ø34
Ø30
Ø76
3 X 45°
102
SECTION A-A
A
A
Ø80
Ø90
Ø30
22
60
20
C
C
Ø80
Ø90
45
B
B
2X R10
Ø86
Ø34
Ø90
D-D
Ø76
SECTION C-C
80
2
12
84
33
3 X 45°
Ø30
Ø34
17
Thickness 2 mm
All sides
SECTION B-B
12
Ø20
10
Ø24
DETAIL D-D
SCALE 2:1

Other useful books by CADIN360

1. 150 CAD Exercises

2. AutoCAD Exercises

3. CAD Exercises

4. 50+ SolidWorks Exercises

5. SolidWorks 200 Exercises

6. Autodesk Inventor Exercises

7. Catia Exercises

8. Siemens NX Exercises

Made in the USA
Monee, IL
09 July 2026